T0192265

Communications
in Computer and Information Science   1841

Rationale

The CCIS series is devoted to the publication of proceedings of computer science conferences. Its aim is to efficiently disseminate original research results in informatics in printed and electronic form. While the focus is on publication of peer-reviewed full papers presenting mature work, inclusion of reviewed short papers reporting on work in progress is welcome, too. Besides globally relevant meetings with internationally representative program committees guaranteeing a strict peer-reviewing and paper selection process, conferences run by societies or of high regional or national relevance are also considered for publication.

Topics

The topical scope of CCIS spans the entire spectrum of informatics ranging from foundational topics in the theory of computing to information and communications science and technology and a broad variety of interdisciplinary application fields.

Information for Volume Editors and Authors

Publication in CCIS is free of charge. No royalties are paid, however, we offer registered conference participants temporary free access to the online version of the conference proceedings on SpringerLink (http://link.springer.com) by means of an http referrer from the conference website and/or a number of complimentary printed copies, as specified in the official acceptance email of the event.

CCIS proceedings can be published in time for distribution at conferences or as post-proceedings, and delivered in the form of printed books and/or electronically as USBs and/or e-content licenses for accessing proceedings at SpringerLink. Furthermore, CCIS proceedings are included in the CCIS electronic book series hosted in the SpringerLink digital library at http://link.springer.com/bookseries/7899. Conferences publishing in CCIS are allowed to use Online Conference Service (OCS) for managing the whole proceedings lifecycle (from submission and reviewing to preparing for publication) free of charge.

Publication process

The language of publication is exclusively English. Authors publishing in CCIS have to sign the Springer CCIS copyright transfer form, however, they are free to use their material published in CCIS for substantially changed, more elaborate subsequent publications elsewhere. For the preparation of the camera-ready papers/files, authors have to strictly adhere to the Springer CCIS Authors' Instructions and are strongly encouraged to use the CCIS LaTeX style files or templates.

Abstracting/Indexing

CCIS is abstracted/indexed in DBLP, Google Scholar, EI-Compendex, Mathematical Reviews, SCImago, Scopus. CCIS volumes are also submitted for the inclusion in ISI Proceedings.

How to start

To start the evaluation of your proposal for inclusion in the CCIS series, please send an e-mail to ccis@springer.com.

Jemal Abawajy · João Manuel R.S. Tavares ·
Latika Kharb · Deepak Chahal · Ali Bou Nassif
Editors

Information, Communication and Computing Technology

8th International Conference, ICICCT 2023
New Delhi, India, May 27, 2023
Revised Selected Papers

 Springer

Editors
Jemal Abawajy
Deakin University
Burwood, VIC, Australia

João Manuel R.S. Tavares ⓘD
University of Porto
Porto, Portugal

Latika Kharb ⓘD
Jagan Institute of Management Studies
Delhi, India

Deepak Chahal ⓘD
Jagan Institute of Management Studies
Delhi, India

Ali Bou Nassif
University of Sharjah
Sharjah, United Arab Emirates

ISSN 1865-0929   ISSN 1865-0937 (electronic)
Communications in Computer and Information Science
ISBN 978-3-031-43837-0   ISBN 978-3-031-43838-7 (eBook)
https://doi.org/10.1007/978-3-031-43838-7

This Springer imprint is published by the registered company Springer Nature Switzerland AG
The registered company address is: Gewerbestrasse 11, 6330 Cham, Switzerland

Paper in this product is recyclable.

Preface

The International Conference on Information, Communication and Computing Technology (ICICCT 2023) was held on May 27, 2023 in New Delhi, India. ICICCT 2023 was organized by the Department of Information Technology, Jagan Institute of Management Studies (JIMS) Rohini, New Delhi, India. The conference received 323 submissions and 60 papers were shortlisted for review and after double-blind reviews and an average of 3 reviews per paper, 14 papers were selected for this volume. The acceptance rate was around 23.3%. The contributions came from diverse areas of information technology categorized into two tracks, namely (1) Intelligent Systems and (2) Pattern Recognition.

The aim of ICICCT 2023 was to provide a global platform for researchers, scientists and practitioners from both academia and industry to present their research and development activities in all aspects of communication and network systems and computational intelligence techniques.

We thank all the members of the Organizing Committee and the Program Committee for their hard work. We are very grateful to Jemal Abawajy, Faculty of Science, Engineering and Built Environment, Deakin University, Australia as General Chair, Ali Bou Nassif, College of Computing and Informatics, University of Sharjah, Sharjah, United Arab Emirates as Program Chair, Oscar Castillo, Tijuana Institute of Technology, Tijuana, Mexico as Program Chair, Saroj Kr. Biswas, Department of Computer Science and Engineering, National Institute of Technology Silchar, Assam, India as session chair for Track 1, and Dilip Singh Sisodia, Department of Computer Science and Engineering, National Institute of Technology Raipur, Chhattisgarh, India as session chair for Track 2.

We thank all the Technical Program Committee members and referees for their constructive and enlightening reviews of the manuscripts. We thank Springer for publishing the proceedings in the Communications in Computer and Information Science (CCIS) series. We thank all the authors and participants for their great contributions that made this conference possible.

July 2023

Latika Kharb
Deepak Chahal

The original version of this book has been revised. The name of editor João Manuel R.S. Tavares had not been displayed correctly. This has been corrected at https://doi.org/10.1007/978-3-031-43838-7_15

Organization

General Chairs

Jemal Abawajy   Deakin University, Australia
Joao Tavares   University of Porto, Portugal

Program Chairs

Ali Bou Nassif   University of Sharjah, United Arab Emirates
Oscar Castillo   Tijuana Institute of Technology, Mexico

Conference Secretariat

Praveen Arora   Jagan Institute of Management Studies, India

Program Committee Chairs

Latika Kharb   Jagan Institute of Management Studies, India
Deepak Chahal   Jagan Institute of Management Studies, India

Session Chair for Track 1

Saroj Kr. Biswas   National Institute of Technology Silchar, India

Session Chair for Track 2

Dilip Singh Sisodia   National Institute of Technology Raipur, India

Technical Program Committee

Anuroop Gaddam   Deakin University, Australia
Shantipriya Parida   Silo AI, Finland
Thanh Thi Nguyen   Deakin University, Australia

Talha Meraj	Comsats University, Pakistan
Bibhu Dash	University of the Cumberlands, USA
Amina Samih	University Abdelmalak Essaadi, Morocco
Fethi Jarray	Gabes University, Tunisia
Shubhnandan S. Jamwal	University of Jammu, India
Abdullah Mohammad Saeed Al Binali	Taibah University, Saudi Arabia
Jacek Izydorczyk	Silesian University of Technology, Poland
Abdel-Badeeh Salem	Ain Shams University, Egypt
Krishna Kumar Mohbey	Central University of Rajasthan, India
Syed Muzamil Basha	Reva University, India
Eser Sert	Malatya Turgut Ozal University, Turkey
Aman Sharma	Jaypee University of Information and Technology, India
Kalpdrum Passi	Laurentian University, Canada
Anoop V. S.	Kerala University of Digital Sciences, India
Azurah A. Samah	University of Malaya, Malaysia
Bui Thanh Hung	Industrial University of Ho Chi Minh City, Vietnam
Ming Cai	Zhejiang University, China
Richard Adeyemi Ikuesan	Zayed University, Abu Dhabi Campus, UAE
Sunil L. Bangare	STES's Sinhgad Academy of Engineering, India
Himanshu Jindal	Jaypee University of Information Technology, India
Yogesh Dandawate	Vishwakarma Institute of Information Technology, India
Malaya Kumar Nath	National Institute of Technology Puducherry, India
Saylee Gharge	Vivekanand Education Society's Institute of Technology, India
Mohammad Shabaz	Model Institute of Engineering and Technology, India
Hima Bindu Maringanti	Maharaja Sriram Chandra Bhanja Deo University, India
Mohan Pratap Pradhan	Sikkim University, India
Khanista Namee	King Mongkut's University of Technology, Thailand
Raman Maini	Punjabi University, India
Jyoti Bhola	Chitkara University, India
Praphula Jain	GLA University, India
Md Zahid Hasan	Daffodil International University, Bangladesh
Shakila Basheer	Princess Nourah Bint Abdulrahman University, Saudi Arabia

Rajneesh Gujral	Maharishi Markandeshwar University, India
Sandeep Sambhaji Udmale	Veermata Jijabai Technological Institute, India
Pushpa Singh	GL Bajaj Institute of Technology & Management, India
Sandhya Bansal	Maharishi Markandeshwar University, India
Astik Biswas	Oracle, India
Vijay Gaikwad	Vishwakarma Institute of Technology, India
Sameerchand Pudaruth	University of Mauritius, Mauritius
Stephen O. Olabiyisi	Ladoke Akintola University of Technology, Nigeria
Kamran Shaukat Dar	University of Newcastle, Australia
B. V. Pawar	North Maharashtra University Jalgaon, India
Samir Patel	Pandit Deendayal Energy University, India
Girish Bekaroo	Middlesex University Mauritius, Mauritius
Manar Alkhatib	Dubai International Academic City, UAE
Ayad Tareq Imam	Isra University, Jordan
Arti Jain	Jaypee Institute of Information Technology, India
Joseph Sefara	Council of Scientific and Industrial Research, India
Martin Puttkammer	North-West University, South Africa
Nadeem Sarwar	Bahria University Lahore Campus, Pakistan
Sabrina Tiun	National University of Malaysia, Malaysia
Dian Saadillah Maylawati	UIN Sunan Gunung Djati Bandung, Indonesia
Sharifah Binti Saon	Universiti Tun Hussein Onn Malaysia, Malaysia
Edwin R. Arboleda	Cavite State University, Philippines
Filippo Vella	Consiglio Nazionale delle Ricerche, Italy
Lamiaa Elrefaei	Benha University, Egypt
Nitin Kumar	National Institute of Technology Uttarakhand, India
Khalid Raza	Jamia Millia Islamia, India
Pradeep Tomar	Gautam Buddha University, India
Arvind Selwal	Central University of Jammu, India
Surjeet Dalal	Amity University, Gurugram, India
Rhowel Dellosa	University of Northern Philippines, Philippines

Contents

Intelligent Systems

Using BERT for Swiss German Sentence Prediction

Oliver Köchli, Philippe Wenk, Christian Zweili, and Thomas Hanne[(✉)] [ID]

University of Applied Sciences and Arts Northwestern Switzerland, Olten, Switzerland
thomas.hanne@fhnw.ch

Abstract. Natural Language Processing builds the foundation for a wide range of language-based applications. While widely spread languages benefit from a wide selection of sophisticated models, lesser known and spoken languages are less supported by specific tools.

In this work, a BERT model was set up to process Swiss German sentences. Besides the fact, that there is no pre-trained model for Swiss German, the diversity of the language leads to significantly worse results than is normally the case with BERT models. Missing processing power, and therefore limited processing time were contributing factors for the unsuccessful model. However, the main reason can be identified in the language itself. Missing grammatical structure as well as a good number of different dialects made it almost impossible for the model to succeed in the language tasks.

To improve the created BERT model, the language database needs to be categorized by dialect. Some efforts are already being made to gather language data in dialects by involving the Swiss people. But since participation is voluntary it may take some time to build a strong database. Additionally, the model needs to be pre-trained separately for each dialect to improve its accuracy and reliability.

Keywords: Natural Language Processing · BERT · Swiss German · Sentence Prediction

1 Introduction

BERT (Bidirectional Encoder Representations from Transformers) is a deep neural network based on a transformer architecture and recommended for natural language processing (NLP). Devlin et al. (2018) from Google (where it is used for query processing) created the network, which available in a pre-trained version for various applications. Unlike previous contextless algorithms such as word2vec or GloVe, BERT considers the surrounding words, which can have an impact on the meaning of the word. It has emerged as a leading application for NLP tasks, along with a few other tools, and has gained a lot of attention in research and practice. The website IEEEXplore.org for instance shows over 2500 results for the keyword BERT. Sixty percent of them date from the last two years. BERT can be used for different NLP tasks such as text classification or even sentence classification, the semantic similarity between pairs of sentences, question answering (Q&A) tasks with paragraph and text summarizations.

J. Abawajy et al. (Eds.): ICICCT 2023, CCIS 1841, pp. 3–15, 2023.
https://doi.org/10.1007/978-3-031-43838-7_1

1.1 Problem Statement

Attempts to predict the next sentence or to fill in a missing word have been made multiple times in English. This is not the case for Swiss German. Suggesting the next word or sentence could improve the speed of writing texts. For example, support employees could benefit when replying to questions from tickets. BERT is a deep neural network that is able to predict the next sentences and fill in a missing word in a sentence. There are different implementation versions of BERT. mBERT focuses on the multilinguality with data from 104 Wikipedia languages (Wu & Dredze 2019a). We have not found any literature about the performance of mBERT in specific domains such as IT, Banking, etc. The language Swiss German is not included in those 104 languages. Due to several reasons mentioned below, we have therefore decided to use BERT with a custom Swiss German data set and not the pre-trained mBERT version:

- mBERT is not pre-trained with a Swiss German data set.
- We do not need the multilinguality (although pretraining with standard German might be useful for Swiss German applications).
- We do not expect that the output would be improved, due to cross-lingual language tasks that may not apply to Swiss German.
- To reduce computing power (only train with a Swiss German data set).

In our literature review, we have not been able to find many papers on using BERT with Swiss German. We would like to use BERT to predict the next sentence for Swiss German sentences. A use case could be to have an integration in MS Teams which predicts next words/sentences and/or fills in missing words. Such an addon for Swiss German has not yet been developed and would be something new that would improve the user messages written in Swiss German. This research could be used and applied also for other dialect languages. The significance for this research will be limited to users who write Swiss German. There are around five million people in Switzerland who define German or Swiss German as their main language (Federal Statistical Office, n.d.). It needs to be considered that the result can vary based on the different Swiss German dialects, as there are quite big differences between them.

1.2 Research Questions

Our main research question is: Is BERT suited for working with Swiss German? This research follows the design science research methodology as, for instance, described by Hevner and Chatterjee (2010). Table 1 shows the five different phases of this methodology. These phases are executed one after the other from top to bottom. The knowledge of the development, evaluation, and conclusion phase flows back into the awareness phase.

Table 1. Design science research phases.

Phase	Research question
Awareness	Were BERT experiments made with Swiss German (Literature review and problem statement)?
Suggestion	Is BERT suited for predicting sentences and masked language modeling in Swiss German?
Development	How does BERT need to be configured to train for Swiss German?
Feedback & Evaluation	Are there significant performance differences between German and Swiss German with BERT? How well is BERT suited?
Discussion	N/A

2   Literature Review

The Bidirectional Encoder Representations Form Transformers, or BERT, is a state-of-the-art language representation model. Instead of reading the given data in one direction (left-to-right or right-to-left) (Peters et al. 2018), BERT has the ability to analyze data bidirectionally, as the name suggests (Delvin 2018). As pre-training has proven to be an effective method to improve natural language processing (NLP) tasks (Rogers et al. 2020), Delvin (2018) has introduced the new framework. The extraordinary high performance in different NLP tasks opened up possibilities in different fields of NLP.

2.1   Power of BERT

Gao et al. (2019) used BERT to classify sentiment through target dependency to automatically read out the sentiment polarity of reviews, news reports or discussions on social media. They conclude that BERT outperforms every other existing deep learning model using a word embedding representation. Tenney et al. (2019) share the sentiment of great performance but clearly state the loss of interpretability. Additionally, they raise the question of whether models like BERT are able to "learn" the abstractions that are important to represent natural language or are "simply modeling complex co-occurrence statistics".

Mozafari et al. (2020) implemented a BERT model to detect hate speech or racial bias in social media. Like others, their fine-tuned model performed better than the base models. Fine-tuning is an important factor, when it comes to the improvement of BERTs capabilities, though it has been stated that there is a limit to the possible improvement (Hao et al. 2020).

BERT owes its power and performance to the large neural network and training on large datasets (Devlin et al. 2018). Sanh et al. (2019) have expressed concerns considering the trend towards bigger models. Many smaller organizations cannot afford the computing power needed to work with BERT. Therefore, they introduced a distilled version of BERT (DistilBERT) with a minimal loss of its capabilities. Another approach to reduce the needed computing power was the introduction of ALBERT. By reducing the parameters of BERT significantly Lan et al. (2019) created the model called A Lite BERT.

2.2 Multilingual BERT

With BERT as a baseline, the model was further developed for cross-lingual tasks. To improve its performance over several languages, multilingual BERT (mBERT) was trained in 104 different languages on raw Wikipedia content (Delvin 2018). Wu and Dredze (2019) have proven mBERT to be exceptionally effective in cross-lingual zero-shot tasks. Even though the performance of the base mBERT is satisfying, they advise using it primarily as a base and fine-tuning it for the development of future multi-lingual NLP models. Pires, Schlinger and Garette (2019) came to similar findings in their research. In their zero-shot experiments, mBERT achieved well over 80% accuracy on part of speech tagging. Additionally, they have found, that mBERT performs better than a monolingual trained model when running the same dataset in the same language. The larger and multilingual repository reduces the dependency on word piece overlap. Wang et al. (2019) made similar discoveries by showing only minimal improvement in bilingual models with an increase in word overlaps. According to Pires et al. (2019), this indicates deeper learning of multilingual representations than vocabulary memorization.

Even though mBERT performs well, Wu and Dredze (2020) discovered a below average performance for the less represented languages of the model. Still, the low resource languages profit from the network and mBERT outperforms monolingual and bilingual models, when compared to the same language. As Swiss German is a low resource language it may be interesting to train and fine-tune mBERT accordingly. On the other hand, Liu et al. (2020) clearly state, that a large dataset is key for creating powerful models.

2.3 The Swiss German Struggle

During our research, we found out, that Manfred Vogel from FHNW is involved in gathering Swiss German data for Training natural language models. Unfortunately, they only collected audio and not written text. After further research, we found one public data set available that we can use for this assignment. This data is referred to in the research question.

The methods and different models mentioned above, perform well when used in their pre-trained languages. While there are several BERT models for dealing with German, English, French and other languages, only a few are able to process Swiss German (Scherrer 2021).

German-trained models have difficulties processing Swiss German, as it differs from German in various linguistic characteristics. Additionally, Swiss German has no orthography standard. These factors render the definition of a standardized structure for sentences difficult. (Hollenstein & Aepli 2015). The main issue was not the inability of a well-trained BERT, as seen in various experiments mentioned above, nor the lack of Swiss German data sets. The inconsistency and number of different dialects of the Swiss German language render the NLP tasks difficult, even for trained models (Scherrer &

Rambow 2010). Nonetheless, Scherrer and Rambow managed to create a model that could create the Swiss German equivalent to a standard German word.

Despite the difficulties of the language, several researchers have dedicated their projects to increase the Swiss German database. The FHNW has launched a project for Swiss German speech recognition using a BERT-like technology called automatic speech recognition (ASR) (Plüss et al. 2020). To train the model, they used audio tracks in Swiss German and translated the sentences into German. Unlike in ASR, the audio corpora used in the FHNW project are not applicable in BERT itself, since the standard BERT relies on written input.

Goldhahn et al. (2012) created, and still do, a large database in different languages. Since 2011 they have been adding Swiss German sentences from Wikipedia to their corpus. The corpora are available in different sizes and may be used for research purposes.

In 2015, Hollenstein and Aepli (2015) published NOAH's corpus with 115'000 manually annotated part-of-speech (PoS) tokens (in this case: verbs, nouns, adjectives, etc.). Their experiments with the BTagger from Gesmundo and Samardzi'c (2012) lead to an accuracy of over 85% in the dialects from Aargau, Basel, Bern, Zurich, and Eastern Switzerland. It remains to be seen how the accuracy will change with a larger repository.

The Institute of Complex Systems (iCoSys) (2021) has created a corpus of over 500'000 Swiss German sentences in cooperation with Swisscom. With their program SWISSCRAWL, they managed to gather sentences with their sources, and the probability that it is an actual Swiss German sentence. To use it with BERT, the dataset will need to be prepared accordingly.

Moreover, the project "Schweizer Dialektsammlung" tries to collect several Swiss German translations for the same Standard German sentences through active engagement of the Swiss people (Swiss Association for Natural Language Processing (SwissNLP) 2021). Even though the issue of inconsistency of the language and dialects might be tackled, the problem of the audio incapability of BERT remains.

2.4 Audio BERT

Chi et al. (2020) have recently introduced an Audio ALBERT (AALBERT) that is able to process audio logs with smaller networks and provides a comparable performance like other pre-trained and larger ASR models. Since ASR models rely heavily on pre-training on large data sets with high-quality transcribed logs (Jiang et al. 2019), models like AALBERT or Mockingjay (A. T. Liu et al. 2020) can deal with non-transcribed audio-logs. Since there is enough data to use a text-based BERT, we will not get more into detail on AALBERT and ASR models.

2.5 Research Gap

NLP – and BERT in particular – is a well-researched area. BERT's exceptionally accurate results lead to significant improvements in text recognition and prediction. Most research is done in widely spread languages like English, German, or Spanish. We have not been able to find research on BERT in Swiss German. There is a model available (mBERT), which according to its developers, would be able to process Swiss German. But since the issue is mostly the inconsistency of the Swiss German language and its dialects, it has not been followed further as of now. Based on the available test data, we will conduct experiments to test our hypothesis.

3 Research Design

With this research, we want to find out if the research questions mentioned in Sect. 1.2 are true. The phases of our design science research approach (see Table 1) are realized as follows: The awareness phase is covered with the literature review that was done in Sect. 2. In the suggestion phase, a concept/research design will be created that describes, how the research will be conducted. In the solution phase, the implementation methodology will be used to set up BERT and the datasets. In the evaluation phase, experiments and usability tests will be conducted. In the suggestion phase, BERT will be set up on a computer, and once the general setup is in place, trained on a Swiss German data set. This will happen in the development phase of this project. Configuration changes will be included in this phase to try to improve the predictability of the Swiss German BERT model. In the last phase, the feedback & evaluation phase testing will take place. Different tests (such as different dialects, domains, sentence complexity, etc.) will be conducted to test the quality of the model.

We will train BERT with the Swiss German language data set we found (Goldhahn et al. 2012). The data that is trained with BERT is divided into training and test datasets that will reflect a systematic variation of independent variables. These variables are described in Sect. 4. The configuration and scripts used in the experiment will be documented in the appendix to validate and repeat the experiments.

4 Experiments

4.1 Architecture Overview and Application Concept of BERT

BERT is powered by a transformer (a mechanism that learns contextual relationships between words in a text). A simple transformer consists of an encoder that reads text input and a decoder that generates a task prediction. BERT just requires the encoder part because its objective is to construct a language representation model. A sequence of tokens is fed into the BERT encoder, which is subsequently transformed into vectors and processed by the neural network. However, before BERT can begin processing, the input must be tweaked and enhanced with additional metadata.

Token embeddings: At the start of the first sentence, a [CLS] token is added to the input word tokens, and at the end of each sentence, a [SEP] token is inserted. The [CLS] stands for Classification (of the sentence) and the [SEP] for separating the sentences.

Embeddings of segments: Each token has a marker that indicates whether it is Sentence A or Sentence B as visible in Fig. 1. In the example sentence, A is represented as "my dog is cute" and Sentence B as "he likes playing". The encoder is thus able to differentiate between the sentences.

Positional embeddings: Each token is given a positional embedding to identify where it belongs in the sentence. As an example, in Fig. 1, the word 'is' has the positional embedding number 3.

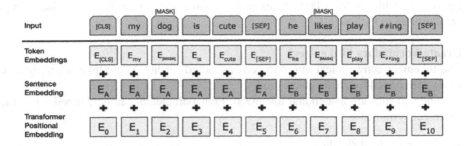

Fig. 1. Different types of embedding for BERT[1].

The experiment needs data for training. As there is no suitable training data for Swiss German, we analyze the performance and output with a pre-trained model data for multilingual application which also supports (standard) German. Also, for Swiss German, we will try to train the model ourselves. With this pretrained version, there will still be some further training or fine-tuning needed before the model can be used. This will take some time and hardware resources. One possibility would be to try using a Google Cloud Instance, as there is some documentation on how to use TensorFlow. Another way is to install TensorFlow on a local machine and use the CPU and GPU for fine-tuning.

Then the predictions on new data can be tested with this fine-tuned model to see if there are valid output.

4.2 Adaptation and Application Scenario of BERT

As mentioned before, as there is no pretrained version of BERT for Swiss German, we will train our own version of BERT. We have followed the tutorial of James Briggs (2021). The following subsections describe how we trained our own version of BERT.

Data Preparation
To train our data model, we will use the following two data sets:

- 100k web from 2017
- 100k Wikipedia from 2021

Both data sets are downloaded from the Uni Leipzig webpage (Uni Leipzig, n.d.). Those two datasets are combined, and new ids are created. A header row is created called "id" and "text", which will be stored as a comma-separated CSV file. The CSV file is then split into a train and a test file after randomly sorting the sentences. The language model should be trained with around 90% of the data. The other 10% is for testing the language model. The data gets loaded into the Python program and then saved into

---

multiple text files with 10'000 entries each. All following steps, including the testing, are executed in Python.

Create Vocabulary (Tokenizer)
The tokenizer converts text into numbers, so-called tokens. These tokens can then be easily interpreted by the language model.

The text files are loaded and tokenized. Four parameters can be defined when the tokenizer is trained:

1. files → the loaded data
2. vocab_size → vocabulary size of the BERT model / number of tokens in the tokenizer (by default 30'522)
3. min_frequency → minimum frequency for a pair of tokens to be merged
4. special_tokens → a list of the tokens that BERT uses, [SEP] or [CLS]

After training the tokenizer, the model is saved.

Prepare for Training
The data is getting loaded by the tokenizer. 15% of the data is now getting masked. Then, the dataset is getting initialized with the masked data. The transformer (our language model) is initialized with the following configuration (explanation of the parameters are provided from Hugging Face (n.d.)):

1. vocab_size = 30_522, → same value as the tokenizer vocab_size
2. max_position_embeddings = 514 → the maximum sequence length that this model might be used with. By default, it is 512. Since the length of our used sentences is rather short, the set parameter will suffice.
3. hidden_size = 768 → Represents the dimensionality of the encoder layers and the pooler layer. By default, it is 768.
4. num_attention_heads = 12 → number of attention heads for each attention layer in the transformer encoder. By default, it is 12.
5. num_hidden_layers = 6 → number of hidden layers in the transformer encoder. By default, it is 12. For our training, we chose 6 hidden layers by following the tutorial.
6. type_vocab_size = 1 → vocabulary size of the token_type_ids passed when calling the model.

Training
The transformer is then trained over two epochs with the prepared dataset. Depending on the hardware, this step can take quite long. With only CPU computational power enabled, this took around 40 min on a desktop PC with an i7-8700k. After the training, we saved the transformer model.

Testing
Now the language model can be tested with single sentences. For each test, one word is masked in the sentence. The language model should fill in an appropriate word. The next section analysis the results in detail.

5   Results

We tested the language model with several sentences. All sentences and the results can be viewed in the appendix. The token is the predicted sentence component. Which means that the token is not limited to a word but can for example be a full stop, space, or nothing at all. The score is the probability (i.e., a value between 0 and 1) that the sentence components fit. A higher score means that the prediction is better. Let us have a detailed look at an example.

Sentence "Ich bin hüt go fill.tokenizer.mask_token."

Responds:

"Ich bin hüt go." (token: ',' score: 0.006835)
"Ich bin hüt go." (token: '.', score: 0.000843)
"Ich bin hüt go er." (token: 'er', score: 0.000467)
"Ich bin hüt go am." (token: 'am', score: 0.000421)
"Ich bin hüt go vorgleit." (token: 'vorgleit', score: 0.000380)

In the sentence above, the last word is masked. This is indicated with the "fill.tokenizer.mask_token" word. This means that the language model needs to predict this sentence component. The language model predicted the five best matching sentence components to be "", ".", "er", "am", "vorgleit". In the given example, any verb in its infinitive form would fit perfectly: "Ich bin hüt go schwimme.", which translates to "I went swimming today". A native-speaking Swiss German person will notice that none of the predicted sentence components make sense. Which is not surprising, as the score for all predictions is below 1%.

If we look at the tests, the accuracy of the model was not very good. Some predictions made the sentence logically correct. But the confidence was always very low. It was always below 1% which represents a very bad confidence rating. The best prediction for each sentence was between 0.55% and 0.68%. Whereas already the second prediction range was between 0.05% and 0.22%. This is a significant loss between the first and the second prediction. The average prediction of 0.18% is, as expected, very low. Jindrich, Rudolf and Alexander (2019) describe the accuracy of mBERT to be 86.7%. The data shows a significant difference in accuracy between our trained model and the pretrained mBERT model.

Considering the tests we conducted, the language model seems to be biased. We have multiple tokens that have been predicted in multiple sentences. These tokens are '', '.', 'er', 'am', 'Garnison'. Table 2 shows the number of occurring sentence components.

We have not found a general rule that says that a certain minimum number of sentences for training is needed. Based on the tests we can conclude that our trained language model with 200'000 sentences is not very accurate.

As pointed out in Sect. 2.3, some of the inaccuracy of our results can be traced back to the inconsistency of the Swiss German language. The different dialects have different words for the same term. This makes it difficult for the model to find the right prediction. This might be countered by treating each dialect as its own "language" and training the model separately for Basel, Zurich, Valais, and so on.

Table 2. Count of sentence components from tests.

Sentence component	Count
' '	6
'. '	6
'er'	5
'am'	5
'Garnison'	4
'vorgleit'	1
'Im'	1
'Dr'	1

6 Conclusion

The model created during this research is not ready to be used with the Swiss German language. The deep neural network model BERT did not perform very well in our tests on Swiss German language tasks. The sentence with the missing word predicted did not make sense in most of the cases, as it is visible in the results in the appendix. As stated initially, the used BERT version was not specifically trained with Swiss German. The confidence score and the results imply that this should be done extensively. We expect to have more accurate results afterwards. Given the limited timeframe as well as the performance limitations of available computing hardware, we have not been able to train the model accordingly. This would be another research topic to invest in intensive training of the BERT model for the intended language.

As elaborated, there are already several quite different BERT implementations. In the scope of this research, we were not able to test multiple versions for Swiss German. Although it would have been an interesting topic to compare the different variations and evaluate them in regard to time, performance, and accuracy of the result.

Swiss German differs very much from the standard German language. Firstly, because there are as many different dialects used in the various cantons in Switzerland. These dialects are already very different in themselves. Secondly, it makes it difficult for the NLP that there is no strict grammatical structure for the dialects. In addition, the same word can be written differently as there is no standardization of spelling. This makes the prediction of missing words more difficult because the breakdown of the sentence is harder. The dataset we used had special sentences because of the web crawling which was utilized to get the data. This complicated the prediction task even further.

Because of the missing data, only masked training was used. One reason for that was that the data set only contained single sentences. Therefore, it was not possible to determine the meaning of a sentence in the context of a paragraph.

As a result, the research question cannot be satisfactorily answered using the given method and the current version of BERT for the Swiss German language. To improve the results, a training model with Swiss German could be developed, other BERT versions could be tested for comparison and another dataset with coherent sentences could be used.

Appendix - Examples of Masked Sentence Predictions

Sentence "Wie gahts fill.tokenizer.mask_token ?".
   Responds:

1. "Wie gahts?" (token: '', score: 0.006162)
2. "Wie gahts.?" (token: '.', score: 0.001763)
3. "Wie gahts er?" (token: 'er', score: 0.000640)
4. "Wie gahts am?" (token: 'am', score: 0.000444)
5. "Wie gahts Garnison?" (token: 'Garnison', score: 0.000354)

   Sentence "S Heidi isch ufem fill.tokenizer.mask_token gstande."
   Responds:

1. "S Heidi isch ufem gstande." (token: '', score: 0.005527)
2. "S Heidi isch ufem. Gstande." (token: '.', score: 0.000452)
3. "S Heidi isch ufem Garnison gstande." (token: 'Garnison', score: 0.000429)
4. "S Heidi isch ufem er gstande." (token: 'er', score: 0.000382)
5. "S Heidi isch ufem am gstande." (token: 'am', score: 0.000320)

   Sentence "Ich bin hüt go fill.tokenizer.mask_token."
   Responds:

1. "Ich bin hüt go." (token: '', score: 0.006835)
2. "Ich bin hüt go." (token: '.', score: 0.000843)
3. "Ich bin hüt go er." (token: 'er', score: 0.000467)
4. "Ich bin hüt go am." (token: 'am', score: 0.000421)
5. "Ich bin hüt go vorgleit." (token: 'vorgleit', score: 0.000380)

   Sentence "fill.tokenizer.mask_token es isch denno e agnämi Abwäggslig."
   Responds:

1. "es isch denno e agnämi Abwäggslig." (token: '', score: 0.005727)
2. "es isch denno e agnämi Abwäggslig." (token: '.', score: 0.002182)
3. "Im es isch denno e agnämi Abwäggslig." (token: 'Im', score: 0.000605)
4. "am es isch denno e agnämi Abwäggslig." (token: ' am', score: 0.000413)
5. "Dr es isch denno e agnämi Abwäggslig." (token: 'Dr', score: 0.000403)

   Sentence "Ganz imene fill.tokenizer.mask_token hinde wirds niemerds me finde."
   Responds:

1. "Ganz imene hinde wirds niemerds me finde." (token: '', score: 0.006576)
2. "Ganz imene. Hinde wirds niemerds me finde." (token: '.', score: 0.001529)
3. "Ganz imene er hinde wirds niemerds me finde." (token: 'er', score: 0.000538)
4. "Ganz imene Garnison hinde wirds niemerds me finde." (token: 'Garnison', score: 0.000435)
5. "Ganz imene am hinde wirds niemerds me finde." (token: 'am', score: 0.000392)

Sentence "S Chuchichästli isch fill.tokenizer.mask_token gsih".
Responds:

1. "S Chuchichästli isch gsih" (token: '', score: 0.006261)
2. "S Chuchichästli isch. Gsih" (token: '.', score: 0.001577)
3. "S Chuchichästli isch er gsih" (token: 'er', score: 0.000687)
4. "S Chuchichästli isch Garnison gsih" (token: 'Garnison', score: 0.000318)
5. "S Chuchichästli isch am gsih" (token: 'am', score: 0.000313)

References

Briggs, J.: Train New BERT Model on Any Language. Towards Data Science (2021). https://tow ardsdatascience.com/how-to-train-a-bert-model-from-scratch-72cfce554fc6

Chi, P.-H., et al.: Audio ALBERT: A Lite BERT for Self-supervised Learning of Audio Representation. arXiv:2005.08575 (2020). http://arxiv.org/abs/2005.08575

Delvin, J.: bert/multilingual.md. (2018). https://github.com/google-research/bert/blob/a9ba4b8d7 704c1ae18d1b28c56c0430d41407eb1/multilingual.md. Accessed 20 Jan 2023

Devlin, J., Chang, M.W., Lee, K., Toutanova, K.: BERT: Pre-training of Deep Bidirectional Transformers for Language Understanding. NAACL HLT 2019 - 2019 Conference of the North American Chapter of the Association for Computational Linguistics: Human Language Technologies - Proceedings of the Conference, 1, pp. 4171–4186, Association for Computational Linguistics (2018). https://doi.org/10.48550/arxiv.1810.04805

Federal Statistical Office. (n.d.). Hauptsprachen seit 1910 - 1910–2020. https://www.bfs.admin. ch/bfs/en/home/statistics/population/languages-religions/languages.assetdetail.20964034. html. Accessed 22 Apr 2022

Gao, Z., Feng, A., Song, X., Wu, X.: Target-dependent sentiment classification with BERT. IEEE Access 7, 154290–154299 (2019). https://doi.org/10.1109/ACCESS.2019.2946594

Gesmundo, A., Samardzic, T.S.: Lemmatisation as a tagging task. In: Proceedings of the 50th Annual Meeting of the Association for Computational Linguistics, pp. 368–372 (2012)

Goldhahn, D., Eckhart, T., Quasthoff, U.: Building large monolingual dictionaries at the Leipzig corpora collection: from 100 to 200 languages. In: Proceedings of the 8th International Language Resources and Evaluation (LREC'12), pp. 759–765. European Language Resources Association (ELRA) (2012) https://wortschatz.uni-leipzig.de/en/download/Swiss% 20German#gsw_wikipedia_2021

Hao, Y., Dong, L., Wei, F., Xu, K.: Investigating learning dynamics of BERT fine-tuning. In: Proceedings of the 1st Conference of the Asia-Pacific Chapter of the Association for Computational Linguistics and the 10th International Joint Conference on Natural Language Processing, pp. 87–92. Association for Computational Linguistics (2020)

Hevner, A., Chatterjee, S.: Design science research in information systems. In: Design Research in Information Systems. Integrated Series in Information Systems, vol. 22, pp. 9–22. Springer, Boston, MA (2010). https://doi.org/10.1007/978-1-4419-5653-8_2

Hollenstein, N., Aepli, N.: A resource for natural language processing of Swiss German dialects. In: International Conference of the German Society for Computational Linguistics and Language Technology (GSCL), pp. 108–109. s.n. (2015). https://doi.org/10.5167/uzh-174601

Hugging Face: Model Documentation Bert. https://huggingface.co/docs/transformers/model_doc/ bert. Accessed 18 June 2022 (n.d.)

Institute of Complex Systems: SWISSCRAWL. Icosys.Ch. (2021). https://icosys.ch/swisscrawl

Jiang, D., et al.: Improving transformer-based speech recognition using unsupervised pre-training. arXiv preprint arXiv:1910.09932 (2019). http://arxiv.org/abs/1910.09932

Wang, Z., Mayhew, S., Roth, D.: Cross-lingual ability of multilingual BERT: An empirical study. arXiv preprint arXiv:1912.07840 (2019). http://arxiv.org/abs/1912.07840

Lan, Z., Chen, M., Goodman, S., Gimpel, K., Sharma, P., Soricut, R.: ALBERT: A Lite BERT for Self-supervised Learning of Language Representations. arXiv preprint arXiv:1909.11942 (2019). http://arxiv.org/abs/1909.11942

Libovický, J.L., Rosa, R., Fraser, A.: How Language-Neutral is Multilingual BERT? (2019). https://github.com/google-research/bert

Liu, A.T., Yang, S., Chi, P.-H., Hsu, P., Lee, H.: Mockingjay: unsupervised speech representation learning with deep bidirectional transformer encoders. In: ICASSP 2020–2020 IEEE International Conference on Acoustics, Speech and Signal Processing (ICASSP), pp. 6419–6423. IEEE, Piscataway (2020) https://doi.org/10.1109/ICASSP40776.2020.9054458

Liu, C.-L., Hsu, T.-Y., Chuang, Y.-S., Lee, H.: What makes multilingual BERT multilingual? arXiv preprint arXiv:2010.10938 (2020). https://doi.org/10.48550/arXiv.2010.10938

Mozafari, M., Farahbakhsh, R., Crespi, N.: Hate speech detection and racial bias mitigation in social media based on BERT model. PLoS ONE **15**(8), e0237861 (2020). https://doi.org/10.1371/journal.pone.0237861

Peters, M.E., et al.: Deep contextualized word representations. In: Proceedings of the 2018 Conference of the North American Chapter of the Association for Computational Linguistics: Human Language Technologies (NAACL-HLT), vol. 1, pp. 2227–2237. Association for Computational Linguistics, New Orleans, Louisiana (2018)

Pires, T., Schlinger, E., Garrette, D.: How multilingual is Multilingual BERT? arXiv preprint arXiv:1906.01502 (2019). https://doi.org/10.48550/arxiv.1906.01502

Plüss, M., Neukom, L., Scheller, C., Vogel, M.: Swiss Parliaments Corpus, an Automatically Aligned Swiss German Speech to Standard German Text Corpus. arXiv preprint arXiv:2010.02810 (2020). http://arxiv.org/abs/2010.02810

Rogers, A., Kovaleva, O., Rumshisky, A.: A primer in Bertology: what we know about how BERT works. Trans. Assoc. Comput. Linguist. **8**, 842–866 (2020). https://doi.org/10.1162/TACL_A_00349/96482/A-PRIMER-IN-BERTOLOGY-WHAT-WE-KNOW-ABOUT-HOW-BERT

Sanh, V., Debut, L., Chaumond, J., Wolf, T.: DistilBERT, a distilled version of BERT: smaller, faster, cheaper and lighter. arXiv preprint arXiv:1910.01108 (2019). http://arxiv.org/abs/1910.01108

Scherrer, Y.: swiss-bert / README.md. Github.Com (9 April 2021). https://github.com/yvesscherrer/swiss-bert/blob/main/README.md

Scherrer, Y., Rambow, O.: Word-based dialect identification with georeferenced rules. In: Proceedings of the 2010 Conference on Empirical Methods in Natural Language Processing, pp. 1151–1161. Association for Computational Linguistics (2010). http://als.wikipedia.org

Swiss Association for Natural Language Processing (SwissNLP): Schweizer Dialektsammlung - Jeder kann mitmachen! (2021). https://www.dialektsammlung.ch/de

Tenney, I., Das, D., & Pavlick, E.: BERT Rediscovers the Classical NLP Pipeline. arXiv preprint arXiv:1905.05950 (2019). http://arxiv.org/abs/1905.05950

Uni Leipzig. Download Corpora Swiss German. (n.d.). https://wortschatz.uni-leipzig.de/en/download/SwissGerman. Aaccessed 1 June 2022

Wu, S., Dredze, M.: Beto, bentz, becas: The surprising cross-lingual effectiveness of BERT. arXiv preprint arXiv:1904.09077 (2019). http://arxiv.org/abs/1904.09077

Wu, S., Dredze, M.: Are all languages created equal in multilingual BERT?. arXiv preprint arXiv:2005.09093 (2020). http://arxiv.org/abs/2005.09093

An ARIMA and XGBoost Model Utilized for Forecasting Municipal Solid Waste Generation

Irfan Javid[1,2(✉)], Rozaida Ghazali[1], Tuba Batool[1], Syed Irteza Hussain Jafri[1], and Abdullah Altaf[1]

[1] Faculty of Computer Science and Information Technology, Universiti Tun Hussein Onn Malaysia, Parit Raja, Malaysia
{irfanjavid, irtezasyed}@upr.edu.pk, rozaida@uthm.edu.my,
hi210007@siswa.uthm.edu.my
[2] Department of Computer Science and IT, University of Poonch Rawalakot, Rawalakot, Pakistan

Abstract. The quantity of urban solid waste continuously increases every year due to a number of variables, including population growth, financial situation, and consumption patterns,. A critical concern for controlling MSW is the lack of basic garbage statistics. To anticipate the total mass of residential trash produced in Multan, Pakistan is the main goal of this study. In the Multan city, 1,901,128 tons of MSW are anticipated to be produced between 2023 and 2027. By utilising machine learning techniques, a particular model has been developed to calculate the overall waste generated quantity for each year. To calculate the quantity of trash generated, the ARIMA and Extreme Gradient Boosting (XGBoost) techniques have been used. In a comparison of the two models, XGBoost outperformed. As a result, the XGBoost's parameters have been adjusted to produce the best results. The hyperparameters that have tweaked into the XGBoost produced the optimum outcomes, with an R^2 coefficient of 0.86 and an RMSE of 736.27. The decision-makers could process and dispose of solid waste more effectively by using the predicted MSW weight. The findings show that a hypertuned XGBoost regressor is the best model, outperforming rest of the candidate models with respect to information criteria, and modified determination coefficient, hence can be used to anticipate the amount of waste generated through the year 2027. This study is intended to assist the Multan Municipal Waste company in designing and planning an efficient waste management system.

1 Introduction

Currently, any product developed to raise the standard of living for people inevitably ends up with a solid waste pile. In spite of the element that polluted trash is a worldwide issue, managing the municipality is the major responsibility of local governments. Municipal Solid Waste (MSW) is a term used to describe waste from sewerages, households, and industries [1]. In its solid or semi-solid state, this waste does not pose any risk to health or environment. The MSW mixture varies significantly from one region to the next. The

J. Abawajy et al. (Eds.): ICICCT 2023, CCIS 1841, pp. 16–28, 2023.
https://doi.org/10.1007/978-3-031-43838-7_2

MSW is also referred to as domestic waste or household waste since it contains a large amount of both soluble and non-soluble solid waste.

The major objective of MSW management is to reduce the volume of solid waste by enhancing resource recovery and reusability. The MSW consists of six basic phases for managing solid waste. They are waste generation, source separation, collection, processing, and conversion, as well as final disposal and transportation [2]. When a material loses its utility, waste is produced. The easiest way to sort waste is right where it is produced because doing so after collection from public bins or house to house would be challenging. The suitable treatment process would be used to the collected garbage in order to recover the raw components. The garbage is then delivered to the proper location for disposal in the end. There is a range of disposal options available, including controlled waste disposal sites, designed dumpsites, thermal treatment of waste, recycling, recycling, and burning [3].

The primary cause for landfilling is MSW. Municipal authorities are responsible for collecting residential trash from the housing region and communal rubbish containers to prevent littering and landfills. The waste is then taken to one of the nearby landfills and dumped there. It ultimately results in inadequate or tainted subsurface water. The World Bank evaluates that about 2.01 billion tons per year waste is generated globally [4]. On average, each person produces between 0.11 and 4.54 kg of trash per day. The amount of waste produced worldwide could double by 2050. It was calculated that the MSW produced 1.6 billion tons of greenhouse emissions. The result would be increased air pollution and global warming. The highest level of MSW production ever recorded in Ireland was in 2007, at an annual rate of 3.4 million tons [25]. It was reduced to 2.6 million tons of MSW in 2014. According to the current trend, MSW is produced at a rate of 14 million tons annually and 2.9 tons per capita [22]. Therefore, inappropriate MSW dumping would have an impact on both the environment and people. Additionally, it was shown that there was a significantly greater correlation between an individual's income rate and their trash generation rate [5].

The safe disposal of MSW is a crucial component of preserving a sustainable ecosystem, according to the WHO [6]. A substantial amount of resources, time, and labor is necessary for the conventional waste management system. Machine learning (ML) approaches have been used as a result to resolve this issue. To measure each model's potential for foretelling the volume of waste generated, a comparison of machine learning models was done [7]. Using the assimilated modeling technique, adaptive optimization based Fuzzy decision support system, the residential trash generation rate was predicted with a high degree of accuracy, reflected in a greater R2 value of 0.87 and a lesser RMSE of 95.7. This study has used the XGBoost regressor and ARIMA modeling technique to forecast the city's MSW generation rate.

2   Related Work

Numerous literary works have been examined to determine the variables, methods, and approaches to anticipate the generation of municipal solid trash, as well as to explain some of the concepts included in the forecast.

A technique for estimating the production of MSW trash in Bangkok by utilizing both Matlab's linear and nonlinear modeling techniques. Population indicators (such as total residents, Native citizens, entire natives between the age of 15–59 years, and all residents between the age of 15–59 years), total municipal solid waste, dwelling indicator (such as the number of homes), economic indicator (such as the average household income), and external indicator (such as the number of tourists) are all thought to be factors that may affect the generation of household trash. In this research, municipal garbage generation has been predicted using ARIMA and XGBoost regressor.

An evaluation of prediction models for municipal solid waste generation included an assessment of a time series forecasting model that utilizes machine learning network for garbage production quantity in Bangkok, with recommendations for the model provided. To perform analysis and build a forecast model, the software tool Rapidminer is utilized, and the ANN model is trained with the backpropagation algorithm.

Time series modeling is a common technique utilized in the analysis of household trash streams and the forecasting of garbage generation in various studies [14]. This study focuses on examining, contrasting, and selecting the optimal time series model for Tehran city to forecast the household trash generation. The outcomes demonstrated that the MAD, MAPE, and RMSE metrics utilized for the prediction of generated MSW amount in the upcoming years have the best performance results by the ARIMA $(2, 1, 0)$ model. To perform simulations and predictions, R software is employed. Additionally covered in the article have been stationary testing, identification of model, diagnostic examination, and scenario reduction.

Since various factors can affect differences in waste output, an accurate estimate of waste production is a crucial first step in developing effective waste management strategies. Predictive and prognostic modelling are helpful tools that provide trustworthy support for decision-making [24, 29]. The prognostic models have used the following indicators to evaluate solid waste quantity as reference attributes: number of residents, status of population, urban residents, and municipal trash quantity. By using the study conducted on Iasi, Romania, the Trash Prognostic Tool, time series analysis, and regression analysis are utilized for anticipating the creation and configuration of MSW. For using the regression models, six types of waste—metal, plastic, glass, compostable trash, paper, and some other kind of garbage—have been identified. The S-curve has been discovered after performance evaluation measures have been computed and results have been obtained.

The studies [9, 23] also covered the topic of creating a modeling technique to precisely estimate MSW quantity that supports municipal authorities in development of a better strategy to run MSW management system. Four intelligent system methods that have been used in this study. The results demonstrated improved forecast effectiveness in artificial intelligent modeling approaches that could be used for constructing successful MSW prediction techniques. Machine learning algorithms [26, 27] are able to accurately anticipate MSW generation on monthly basis by training on time series generation of garbage. The outcomes also show that KNN predicted trash monthly averages, however ANFIS structures generated the most accurate peak forecasts.

To estimate the MSW on weekly basis for network-level, Kontokosta et al. used the cognitive network and Stochastic Boosting Forests modeling techniques. With an R^2 value of 0.87, the GBRT performed best [10]. To anticipate the production of MSW, a model using decision trees and ANNs has been developed [11]. According to predictions made by Kannangara et al., ANN has an accuracy rate of 72%, a minor MAPE of 16% and a greater coefficient of correlation value (0.72). To forecast home garbage, Navarro-Esbri et al. utlized a temporal evaluation. The SARIMA model was used to estimate the monthly and daily predictions of MSW.

The management of MSW is a significant subject of research in recent years. By 2050, it was predicted that there will be 27 billion tons of trash generated per year worldwide [12]. Asia would be responsible for one-third of the world's waste, with China, India, and Pakistan making the largest contributions. MSW disposal is at a dangerous stage right now, thus it's imperative to take action and provide services to handle the growing volumes of MSW. This research suggests a modeling technique to calculate the MSW quantity, which will aid municipal authorities in reaching the best decisions.

3 Material and Methods

The core phases of the approach that has been applied in this research are shown in Fig. 1. The Actual data has been preprocessed, and the ARIMA model and XGBoost forecasting models, whose forecasting outcomes would be further compared and discussed, are optimized and executed on the basis of same dataset.

3.1 Study Area Information

Multan is a big city located in South Pakistan and is a well-developed city because of its bigger capacity in industrial development and modernization [13]. Multan has now approximately 2.16 million population of permanent residents that leads to augmentation of generated MSW in Multan, reaching 1.16 million tons in 2021 [15]. The management of MSW in Multan includes collection, transportation, and disposal together. Owing to the under privileged MSW management and swift growth of MSW generation, the disposal options in Multan are going through trouble for satisfying the MSW treatment requirements in recent years [16]. Therefore, Multan authorities provides services of MSW collection, transfer, and disposal to 68 union council of Multan. The city Multan has been chosen as the area of study for catering MSW prediction results and recommendations to the municipal administration.

3.2 Data Acquisition and Preparation

The waste management company (WMC) in Multan, which is in charge to collect, clean, transport, and dispose MSW for management in the city's central business district, provided the information that has been utilized in this research. The provided information includes generation of MSW in 68 union councils of Multan. Weekly original statistical data collection took place between July 1, 2017, and June 30, 2022, totaling 782 data samples. The data was partitioned into two segments, where 80% was allocated to the training dataset and the remaining 20% was allocated to the testing dataset. During the training phase of the ARIMA and XGBoost, the training dataset has been further split into two halves, one for training the model and the other for cross-validation.

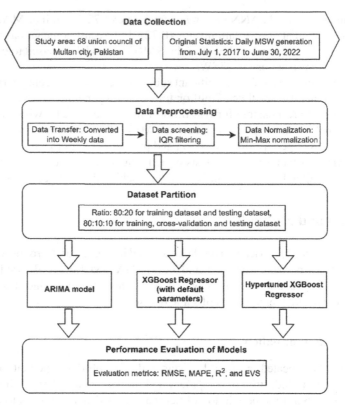

Fig. 1. An overview of methodology.

During the collection of the data and its several dimensions, random error occurs that result in significant statistical differences, data pre-processing has been done for the removal of erroneous data sections and dimensions impact elimination [17]. In this research, the main data preprocessing stages are data transfer, screening, and normalization. By summing up the generation of MSW over the course of seven successive days, actual daily base data has been transformed into weekly basis statistics by taking into account the weekly work system. Interquartile range (IQR) filtering has been employed to screen the data for preventing any detrimental impacts brought on by substantially deviated data.

The outliers have been identified through IQR filtering by dividing the dataset into equal quarters that relies on each data section value, the Eqs. (1), (2), and (3) are used to calculate the dataset's IQR, and identifying the higher and lower bounds of the usable dataset.

$$IQR = R_3 - R_1 \qquad (1)$$

$$lower\ quartile = R_1 + \left(\frac{N}{4} - \frac{f_C}{f_Q} \right) * w \qquad (2)$$

$$upper\ quartile = R_3 + \left(\frac{3N}{4} - \frac{f_C}{f_Q}\right) * w \tag{3}$$

where R_1 and R_3 are lower and upper quartile, N is the distribution's overall frequency, f_c represents the accumulative frequency before the quartile class, f_Q represents the frequency of the quartile class, and w represents the width of the quartile class.

After removing invalid data, there are 532 and 77 data samples available for the training and testing datasets, respectively. The application of min-max scaling was carried out to eradicate the influence of measurement. Transforming data samples into values between 0 and 1, min-max normalization is applied by their size proportionality in the initial dataset. This process is depicted in Eq. (4).

$$x_{norm} = \frac{x - x_{min}}{x_{max} - x_{min}} \tag{4}$$

where x_{norm} is the data after normalization, x is the actual data, and x_{min} and x_{max} are the least and supreme samples in the actual dataset.

3.3 XGBoost

Ensembling of a decision tree that relies on gradient boosting with an objective of being highly scalable is known as XGBoost. XGBoost is primarily responsible to construct a stabilizer expansion in the objective function via minimising a loss function, much to gradient boosting. The complexity of the trees has been regulated by using a different loss function because of the reason that XGBoost merely uses decision trees for being its basic classifiers.

$$\mathcal{L}(\emptyset) = \sum_i l(\hat{y}_i, y_i) + \sum_k \Omega(f_k)$$
$$where\ \Omega(f) = \gamma T + \frac{1}{2}\lambda \|w\|^2 \tag{5}$$

The quasi residuals of the forecasted results y_i with hat and the actual data value y_i in every single leaf are calculated in the first portion of the above mentioned equation, which represents the loss function. The equation's second portion combines the two sections just mentioned before. The "regularization term lambda" in the final component of the omega formula is meant to "lower the prediction's insensitivity to individual data," and "w" stands for the leaf weights, which we can also consider as the leaf's output value. Additionally, T stands for the quantity of the ending nodes or leaves in a tree, while gamma is the forfeit that is user-definable and intended to promote pruning.

Additionally, randomization approaches have been utilizes by XGBoost with the purpose of reducing training time and eliminate over-fitting. XGBoost has incorporated column subsampling at the tree and tree node levels and random subsamples for the training of individual trees as randomization approaches. Moreover, XGBoost has employed number of different methods to speed up the training phase in decision tree, which are not exactly linked to the ensemble accuracy. The determination of the best split is the most onerous phase in the construction of decision tree algorithms, where XGBoost has majorly focused on reducing the computational complexity. Most of the algorithms used for the discovery of best splits, come up with the list of all splits that are potential candidates prior to the selection of the one split that has largest gain. Running of a linear scan for every single sorted feature is necessary to determine the optimum split for every node [28]. XGBoost has used a unique compressed column-based format where data has to be pre-sorted and stored for the prevention of repetition in data sorting for every node. This is the way XGBoost manages to sort every single attribute just once. The determination of an ideal split for every single feature which is under consideration becomes more expedient through this column-based storage structure.

3.4 ARIMA

According to the study [8], the ARIMA model is a fusion of the moving average, autoregression, and differencing techniques. One can fine-tune the ARIMA model's parameters, namely the number of auto-regressions (p), order of differencing (d), and the length of the moving average (q), as indicated in Eq. (6) [18], ARIMA provides the forecasted value y_t for time t:

$$y_t = \mu + \sum_{i=1}^{p} \beta_i y_{t-i} + \varepsilon_t + \sum_{i=1}^{q} \theta_i \varepsilon_{t-i} \tag{6}$$

where μ represents a constant, β_i denotes to a partial correlation coefficient of y_{t-i}, and θ_i are representing the parameter coefficient of white noise ε_{t-i}.

To use the ARIMA model, the sequence needs to undergo some modifications so that it could have an established probabilistic features and only varies within the bounds of the series' constant mean and variance [19]. New sequences are continually generating during the smoothing process by the mean of computing the variances that lies in the sequential data in the prior sequence unless the sequence turns out to be stationary, parameter d is determined.

4   Experiments and Results

The municipal solid trash (measured in tons) distribution has been done according to the financial quarter, as depicted in Fig. 2. As shown in Fig. 3, the dataset has been divided into 20% of testing data and 80% of training data. The ARIMA and XGBoost regressor have been used for the prediction of the waste amount because it is a regression problem. Here, RMSE, MAPE, R^2, and Explained Variance Score have been used as Key Performance Indicators (KPI).

- Root Mean Square Error (RMSE)
    The quantifiable data has predicted with errors, which are calibrated by RMSE. In order to identify the residues, it is used to compute the distance that lies in the data points and the regression line.

$$RMSE = \sqrt{\frac{1}{n} \sum_{i=1}^{n} \frac{(x_i - y_i)^2}{n}} \tag{7}$$

where, $x_1, x_2, \ldots x_n$ are the observed values, $y_1, y_2, \ldots y_n$ represents the forecasted data values, and 'n' represents the sample's quantity.

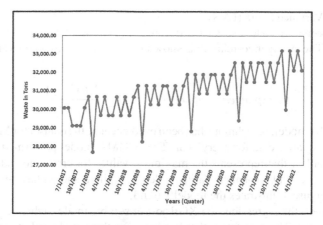

Fig. 2. Availability of Data.

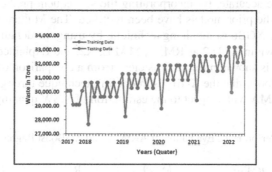

Fig. 3. Division of data into Training and Testing sets.

- Mean Absolute Percentage Error (MAPE)

  The method has been employed to gauge the effectiveness of the approach in forecasting the system, for datasets of both large and small sizes. MAPE offers the best fit.

$$MAPE = \frac{1}{n} \sum \frac{A-F}{A} *100 \tag{8}$$

- Coefficient of determination (R^2)

  It is the main outcome of a regression study. It is calculated by squaring the correlation (r), which varies from 0 to 1, between the actual and predicted y values.

$$r = \frac{m\left(\sum ab\right) - \left(\sum a\right)\left(\sum b\right)}{\sqrt{\left[m\sum a^2 - \left(\sum a\right)^2\right]\left[m\sum b^2 - \left(\sum b\right)^2\right]}} \tag{9}$$

- Explained Variance Score (EVS)

It has been utilized to assess the difference that lies among the data points and the prototype. The stronger correlation is shown by the larger Explained Variance Score percentage.

$$explained\ variance(y, \hat{y}) = 1 - \frac{Var((y, \hat{y}))}{Var(y)} \qquad (10)$$

The ARIMA modeling technique has been used because the dataset for Multan-MSW consists of the quarter data for every year. The ARIMA model performed mediocrely, as shown in Table 1 that represents the maximum values for error rate and least values for the value of R^2 and Explained Variance Score, while it should have been the other way around. Figure 4 illustrates the model's trend.

In order to do the regression, the XGBoost ensemble model, which is quite versatile, has been used. The correlation that lies between the actual attribute and the target attribute is produced by this ensemble model by repeatedly adding models until the forecasted facts are accurate. By incorporating the subsequent predictors, the residues issue triggered in the prior models have been rectified. The Multan-MSW dataset has been applied to the XGBoost modleing technique by using the default parameters, and the results are shown in Table 2 as RMSE, MAPE, R2, and Explained Variance Score. The model's trend is shown in Fig. 5. It is clear from a comparison of the two models, ARIMA and XGBoost, that the XGBoost modeling technique through default settings outperformed ARIMA with respect to the estimation of MSW quantity.

Table 1. Performance Evaluation of ARIMA and XGBoost Regressor model

Model	RMSE	MAPE	R^2	EVS
ARIMA	2218.76	274.05	0.10	−0.02938
XGBoost	845.07	124.36	0.74	0.8615

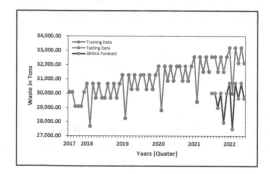

Fig. 4. MSW Forecasting using ARIMA Model.

Although the XGBoost achieved superior outcomes, the parameters have been optimized to enhance the model's performance. Max_depth and n_estimator, two booster

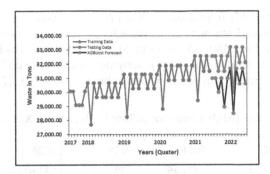

Fig. 5. MSW Forecasting by XGBoost Regressor with default parameters

parameters, have been hyperparameterized. A tree (max_depth) parameter's value with the maximum depth causes the model to grow more complex. Therefore, the max_depth size has to be decreased to prevent overfitting. The Number of gradients boosted trees, or n_estimators, has a default value of 100. A series of values from 50 to 350 with an interval of 50 could be used to evaluate the hyperparameter n_estimator [20]. The XGBoost model produced the best results by setting parameters max_depth and n_estimator to 2 and 400, respectively. The tendency of the hyperparameterized XGBoost model is shown in Fig. 6. By yielding the RMSE, MAPE, R^2, and Explained Variance Scores of 736.27, 122.96, 0.86, and 0.8741, respectively, as shown in Table 2, this technique attains the optimum results as compared to the default model.

Table 2. Performance comparison of XGBoost and Hypertuned XGBoost Regressor model

Model	RMSE	MAPE	R^2	EVS
XGBoost (default parameter)	845.07	124.36	0.74	0.8615
XGBoost (with hyperparaterers)	736.27	122.96	0.86	0.8741

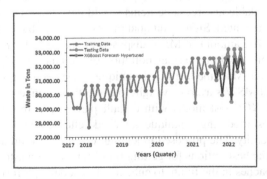

Fig. 6. MSW Prediction by Hypertuned XGBoost Regressor

In Table 3, the monthly forecasted MSW generation outcomes using a hyperparameterized XGBoost model are presented. According to the prediction, annual garbage production will drop from 380,245 tonnes in 2023 to 380,208 tonnes in 2027. This can be attributed to the rise in informal sector activities, such as the collecting of rubbish and recycling in the city by scavengers who visit homes and streets [21].

Table 3. Predicted MSW Generation in Tons by Hypertuned XGBoost Model

Month	2023	2024	2025	2026	2027
January	32,627	32,626	32,625	32,624	32,623
February	29,470	29,469	29,469	29,468	29,467
March	32,627	32,626	32,625	32,624	32,623
April	31,575	31,574	31,573	31,572	31,571
May	32,627	32,626	32,626	32,625	32,624
June	31,574	31,573	31,573	31,572	31,571
July	31,961	31,960	31,959	31,958	31,957
August	31,953	31,952	31,953	31,951	31,951
September	30,942	30,941	30,940	30,941	30,940
October	31,974	31,973	31,972	31,971	31,970
November	30,942	30,941	30,940	30,941	30,940
December	31,973	31,972	31,971	31,971	31,971
Total	**380,245**	**380,233**	**380,226**	**380,218**	**380,208**

5   Conclusion and Future Work

The MSW computation would increase public knowledge of the amount of waste generated daily per person, which would help people understand the state of the environment today. Utilizing recyclable and reusable items would have the effect of decreasing the amount of garbage generated. Significant quantities of solid garbage are still dumped in landfills in Multan. The amount of MSW disposed of significantly increase during the millennium era. There would be serious pollution of the land, the air, and the water if this situation persisted. Machine learning techniques have been employed for the forecast of the generated MSW's average weight to avoid these concerns. When compared to the ARIMA and XGBoost model with default parameters, the proposed hyperparameterized XGBoost modeling technique achieved better results for the estimation of municipal waste quantity. The proposed modeling technique would assist the administrators in selecting the best options for getting rid of municipal waste. We intend to use deep learning approaches in the future to enhance the model's performance.

Acknowledgment. This research was supported by Ministry of Higher Education Malaysia (MOHE) through the Fundamental Research Grant Scheme (FRGS/1/2020/ICT02/UTHM/03/5).

References

1. Fadhullah, W., Imran, N.I.N., Ismail, S.N.S., et al.: Household solid waste management practices and perceptions among residents in the East Coast of Malaysia. BMC Public Health **22**, 1 (2022). https://doi.org/10.1186/s12889-021-12274-7
2. Ahsan, M., Alamgir, M., El-Sergany, M., Shams, S., Rowshon, M.K., Nik, N.N.: Daud, assessment of municipal solid waste management system in a developing country. Chin. J. Eng. 11, 561935 (2014). https://doi.org/10.1155/2014/561935
3. Abdel-Shafy, H.I., Mansour, M.S.: Solid waste issue: Sources, composition, disposal, recycling, and valorization. Egypt. J. Petrol. **27**(4), 1275–1290 (2018). ISSN 1110–0621, https://doi.org/10.1016/j.ejpe.2018.07.003
4. Jayaraman, V., Parthasarathy, S., Lakshminarayanan, A.R., Singh, H.K.: Predicting the quantity of municipal solid waste using XGBoost model. In: 2021 Third International Conference on Inventive Research in Computing Applications (ICIRCA), Coimbatore, India, pp. 148–152 (2021). https://doi.org/10.1109/ICIRCA51532.2021.9544094
5. van Beukering, P., et al.: Analysing Urban Solid Waste in Developing Counttries: a Perspective on Bangalore (India) (1999)
6. Tanaka, M.: Sustainable Society and Municipal Solid Waste Management (2014). https://doi.org/10.1007/978-981-4451-73-4_1
7. Soni, U., Roy, A., Verma, A., Jain, V.: Forecasting municipal solid waste generation using artificial intelligence models—a case study in India. SN Appl. Sci. **1**(2), 1–10 (2019)
8. Vochozka, M., Horák, J., Šuleř, P.: Equalizing seasonal time series using artificial neural networks in predicting the Euroeyuan exchange rate. J. Risk Financ. Manag. **12**(2) (2019)
9. Abbasi, M., El Hanandeh, A.: Forecasting municipal solid waste generation using artificial intelligence modeling approaches. Waste Manag. **56**, 13–22 (2016). https://doi.org/10.1016/j.wasman.2016.05.018
10. Viljanen, M., Meijerink, L., Zwakhals, L., et al.: A machine learning approach to small area estimation: predicting the health, housing and well-being of the population of Netherlands. Int. J. Health Geogr. **21**, 4 (2022). https://doi.org/10.1186/s12942-022-00304-5
11. Kannangara, M., Dua, R., Ahmadi, L., Bensebaa, F.: Modeling and prediction of regional municipal solid waste generation and diversion in Canada using machine learning approaches. Waste Manag. **74**, 3–15 (2018)
12. David, A., Thangavel, Y.D., Sankriti, R.: Recover, recycle and reuse: an efficient way to reduce the waste. Int. J. Mech. Prod. Eng. Res. Dev **9**, 31–42 (2019)
13. Nadeem, M., et al.: Exploring the urban form and compactness: a case study of Multan, Pakistan. Sustainability **14**(23), 16066 (2022). https://doi.org/10.3390/su142316066
14. Marandi, F., Ghomi, S.: Time series forecasting and analysis of municipal solid waste generation in Tehran city. In: 2016 12th International Conference on Industrial Engineering (ICIE), pp. 14-18 (2016). https://doi.org/10.1109/induseng.2016.7519343
15. Murtaza, G., Habib, R., Shan, A., Sardar, K., Rasool, F., Javeed, T.: Municipal solid waste and its relation with groundwater contamination in Multan, Pakistan. IJAR **3**(4), 434-441 (2017)
16. Shoaib, M., Sarwar, M.: Review and status of solid waste management practices in Multan, Pakistan. Electronic Green J. **1** (2006). https://doi.org/10.5070/G312410671
17. Mojtaba, J., Rahimullah, N.: Prediction of municipal solid waste generation by use of artificial neural network: a case study of Mashhad. Int. J. Environ. Res. **2**(1), 13e22 (2007)
18. Hikichi, S.E., Salgado, E.G., Beijo, L.A.: Forecasting number of ISO 14001 certifications in the Americas using ARIMA models. J. Clean. Prod. **147**, 242e253 (2017)
19. Box, G.E.P., Jenkins, G.M., Reinsel, G.C.: Time Series Analysis: Forecasting and Control, 4th edn. John Wiley & Sons Inc, Hoboken, NJ (2008)

20. https://www.analyticsvidhya.com/blog/2016/03/complete-guide-parameter-tuning-xgboost-with-codes-python/
21. Chibueze, T., Naveen, B.P.: Activities of informal recycling sector in North-Central, Nigeria. Energy Nexus **1**, 1–7 (2021)
22. https://www.epa.ie/irelandsenvironment/waste/
23. Wahid, F., Ismail, L.H., Ghazali, R., Aamir, M.: An efficient artificial intelligence hybrid approach for energy management in intelligent buildings. KSII Trans. Internet Inf. Syst. **13**(12), 5904–5927 (2019). https://doi.org/10.3837/tiis.2019.12.007
24. Zulqarnain, M., Ghazali, R., Mazwin, Y., Rehan, M.:A comparative review on deep learning models for text classification. Indonesian J. Electr. Eng. Comput. Sci. **19** (2020). https://doi.org/10.11591/ijeecs.v19.i1.pp325-335
25. https://www.cso.ie/en/releasesandpublications/ep/p-eii/eii18/waste/
26. Shah, H., Ghazali, R.: Prediction of earthquake magnitude by an improved ABC-MLP. In: 2011 Developments in E-systems Engineering, Dubai, United Arab Emirates, pp. 312–317 (2011). https://doi.org/10.1109/DeSE.2011.37
27. Javid, I., Ghazali, R., Syed, I., Zulqarnain, M., Husaini, N.A.: Study on the Pakistan stock market using a new stock crisis prediction method. PLoS ONE **17**(10), e0275022 (2022). https://doi.org/10.1371/journal.pone.0275022
28. Javid, I., et al.: Optimally organized GRU-deep learning model with chi2 feature selection for heart disease prediction. J. Intel. Fuzzy Syst. **42**(4), 4083–4094(2022)
29. Javid, I., et al.: Data Pre-processing for Cardiovascular Disease Classification: A Systematic Literature Review. 1525 – 1545 (2023)

An Empirical Study of Intrusion Detection by Combining Clustering and Classification Methods

Remah Younisse[ID], Yasmeen Alslman[ID], Eman Alnagi[ID],
and Mohammad Azzeh[(✉)][ID]

Princess Sumaya University for Technology, Amman, Jordan
{rem20219007,yas20219006,ema20219005}@std.psut.edu.jo,
m.azzeh@psut.edu.jo

Abstract. Intrusion detection applications are becoming more critical, and the need for more advanced intrusion detection systems is increasing. Many existing machine learning and artificial intelligence applications are used for this task. Nevertheless, these models were claimed to operate better when the data was clustered before the classification process. In this work, we study the effect of clustering datasets related to attacks applied in different network environments on the classification results generated by many famous and repeatedly used machine learning models. Two different clustering methods were analyzed in this work combined with nine machine learning models; After two hundred and forty-three experiments, it has been concluded that using DBSCAN clustering before classification can slightly enhance the overall model performance. Nonetheless, the enhancement did not exceed 1% in the field of intrusion detection systems.

Keywords: Intrusion detection systems · Network Security · Clustering and classification · Support Vector Machines · Neural Networks · Cross-validation · Unsupervised learning and clustering · Supervised learning by classification · Ensemble methods · Cluster analysis · Data cleaning · Feature selection

1 Introduction

Intrusion detection systems (IDS) are developed to ensure computer network security against malicious and suspicious behaviors [1]. Creating efficient IDS is becoming essential, especially with the increase in cloud usage and the spread of internet of things (IoT) systems [10]. The evolution in machine learning (ML) was reflected in IDS as many advanced and efficient systems were developed [24]. Supervised and unsupervised models were proposed and used extensively with IDS. Supervised models include Naïve Bayes, Classification and Regression, Decision Trees, Logistic Regression, Neural Networks, and many others. While unsupervised models used in literature are usually clustering and Variational Autoencoder models [22]. Using unsupervised learning techniques for intrusion detection was mentioned in [12] as an efficient method for unlabeled data. Semi- Supervised methods were also mentioned as methods combining labeled with

© The Author(s), under exclusive license to Springer Nature Switzerland AG 2023
J. Abawajy et al. (Eds.): ICICCT 2023, CCIS 1841, pp. 29–45, 2023.
https://doi.org/10.1007/978-3-031-43838-7_3

unlabeled data in the model to create an intrusion detection system. In this paper, we focus on utilizing clustering as a preparation step for the data before using it to train supervised learning models, as has been tackled in previous work such as [3].

Clustering methods are well-known as unsupervised learning models usually used with unlabeled data. They aim to gather similar objects in the same groups. Distance-based clustering methods such as k-means are popular approaches that start with selecting k centroids. Then, it computes the distance between data instances and these centroids. Every data instant joins the group whose centroid is the nearest, evaluated by Euclidean distance measures. Then the mean value for each group (cluster) is calculated to become its new centroid. Then the data is clustered again, iteratively, and every data instant joins the group whose centroid is the closest; the process repeats until convergence [5]. On the other hand, density-based clustering focus on grouping the data samples based on density. It targets identifying clusters of various shapes and sizes with different densities [8].

IDS applications extensively use classification algorithms. In literature, [6, 19, 20] are examples of such work. The work in [19] mentioned that the Random Forrest model is considered a very robust classifier in the context of IDS applications. The work also mentioned that combining multiple classifiers can improve IDS classifying applications.

In literature, many works focused on the benefits of merging data clustering approaches with classification techniques, mainly to improve the classification models' performance. This paper presents an empirical study to analyze the effect of applying clustering to intrusion detection datasets before the classification process. Two different clustering algorithms were applied to the used datasets; the clustering result was then appended to the dataset as a new feature. Then the profit of adding this feature was measured on various machine learning models. In Sect. 2, we present the related works in this context. While in Sect. 1, the methodology followed through this study is presented, the used clustering and classification methods and the datasets are discussed in Sect. 1. The experiment results are illustrated and discussed in Sect. 4. Finally, the paper is concluded in Sect. 5.

2   Related Work

Data classification has been extensively implemented with different types of Machine Learning (ML) and Deep Learning (DL) algorithms to apply prediction or detection tasks. Classification is considered a supervised learning technique since datasets are already labeled with certain classes related to the domain under study. As an unsupervised method, clustering is usually used when the data is unlabeled. Thus, clustering will help in grouping similar instances into separate groups and thus allow experts to label each group according to the domain.

In recent studies, clustering and classifications are combined to enhance the classification model by either using clustering as a labeling technique or taking advantage of the resulting cluster labels in the classification task. Table 1 summarizes the papers reviewed in this section.

Classification and clustering techniques are used to apply the desired tasks using datasets from different domains. Management and Finance [15, 17], health [9, 11, 21], security [14], education [13], and humanity studies [7] are just a few examples of domains that data can be extracted from and put under study. Nevertheless, some literature concentrated on the usage of clustering and classifications algorithms without taking into consideration a certain domain [3, 4, 18] As clustering methods, k-means is the dominant method used in most surveyed work, density-based clustering techniques, such as DBSCAN, are not used that intensively. We used k-means and DBSCAN in this empirical study for the clustering step to identify the best to use in our case. As for classification techniques, ML methods are the most dominant in the reviewed literature.

As aforementioned, clustering is usually used to group similar unlabeled data. As a result, each instance is given a cluster ID that is then used as a label for the classification step. Authors of [4] have used k-mode in order to cluster the unlabeled datasets, then used Logistic Regression (LR), Random Forest (RF), and Naive Bayes (NB) as classifiers to train the datasets.

In [14], another clustering method is used, GMM (Gaussian mixture models), also for labeling the datasets before applying an ensemble deep neural network to detect power theft automatically. [17] used k-means for the same purpose, labeling, and used several types of classifiers for training, such as; K-nearest Neighbors (KNN), Support Vector Machine (SVM), Decision Tree (DT), Cat- boost, and Ada. They have used these classifiers to predict the performance of managers in local companies.

Another field of research that needed the clustering step for data labeling is [9], where k-means is used for this task, and an ensemble model using XGBoost (eXtreme gradient boosting) has been used for classification. The classification task here has been conducted in order to predict the risk factors that might occur to patients with a breast cancer history and may have the potential for cancer to return.

The rest of the literature reviewed in this section has used clustering methods to generate clusters for the datasets and thus uses the resulting clusters as new features to be augmented to the dataset. Then, the classifiers will be applied to the new datasets.

Researchers in [3] and [18] have conducted their study on datasets with general domains using k-means as a clustering method. In addition to k-means, [3] have applied another hierarchical clustering method, while [18] used Affinity Propagation. The resulting cluster ids have been added to the datasets in both cases. As for the classification method, [3] used NB and NN, while [18] used SVM, RF, KNN, DT, and other classifiers and conducted a comparison between their results. They have shown that the accuracy has been enhanced due to the usage of the clusters as features, but only for some classifiers.

In [21], similar approach has been conducted on health care datasets. Researchers of [21] have studied the prediction of Parkinson's disease, while researchers of [11] have concentrated on the prediction of Gestational Diabetes, which occurs when high blood sugar rates are detected for women during pregnancy. Such sensitive applications need to yield high accuracy in prediction. Thus, both have used k-means to cluster the data. Although in [21], the data has already been labeled, they have conducted the clustering step to highlight the similarities between instances that may be missed if only classification methods are used. As for [11], they have conducted several k-means clustering trials to find the best k (number of resulting clusters) to enhance the classification step. For classification, [21] used Siamese Neural Network, while [11] used several classification methods; DT, RF, SVM, KNN, LR, and NB as an ensemble model.

Authors of [15] have worked on product classification for e-commerce applications. They have tested several combinations of clustering and classification methods. They found that the best combination for their dataset was to use k-means for data clustering and then use the resulting clusters as a feature to be added to the dataset. In their results, they have not noticed an apparent enhancement in classification. Nevertheless, the execution time of the KNN classifier has been enhanced when adding the resulting clusters to the dataset.

Humanity studies and Education had their share in related work. As in [7], their study concentrated on the prediction of healthy aging of citizens. They used hierarchical clustering and added the resulting clusters to the dataset. An ensemble model classifier has been used for prediction.

As for [13], they have studied the learning styles of students from several perspectives. They have found that students may have several learning styles, not only one. So, they have started with k-means methods to cluster their data. The resulting cluster number for each instance is added as a new feature to the dataset. SVM has been used to train the dataset to predict each student's learning style. Nevertheless, since there can be a combination of learning styles used by each student, they have extended their experiments by using DT to predict these combinations by considering the slight similarities between the different clusters.

Table 1. Related Work Summary

Ref#	Domain	Clustering Method	Classification Model	Clustering Role
[4] 2022	General	K mode	LR, RF and NB	Automatic Labeling
[14] 2020	Power Theft Detection	GMM	Ensemble DNN	Clustering phase is used for data tagging
[17] 2022	Managerial Performance Detection	K-means	KNN, SVM, DT, Catboost, ada	Clustering phase is used for data labelling
[9] 2019	Risk Factors Prediction for Breast Cancer patients	K-means	XGBoost	Clustering phase is used for data labelling

(*continued*)

Table 1. (*continued*)

Ref#	Domain	Clustering Method	Classification Model	Clustering Role
[3] 2016	General	K-means and Hierarchical Clustering	NB and NN	Adding cluster id to the data set to enhance the classification results
[18] 2021	General	K-meansand Affinity Propagation	DT, KNN, RF, SVM and others	Using resulting clusters in data augmentation before classification
[15] 2018	Product Classification (E-commerce)	K-means and Hierarchical Clustering	NB, SVM, KNN and RF	Clustering before classification has reduc ed execution time
[7] 2021	HealthyAging Prediction (Humanity Studies)	Hierarchical Clustering	Ensemble Model	Divide the dataset into k datasets (clusters), and then apply ensemble model
[21] 2021	Parkinson's Disease Classification (Medical)	K-means	Siamese Neural Network	Clustering before classification, to highlight similarities
[11] 2022	Gestational Diabetes Prediction (Medical)	K-means	Ensemble Model (DT, RF, SVM, KNN, LR, and NB)	Applying clustering for data reduction before classification
[13] 2019	Students Learning Style Prediction (Education)	K-means	SVM and DT	Cluster number is added as a feature in the dataset before classification

3   Methodology

Throughout this study, we use three intrusion detection datasets: the ARP- related data from the Kitsune dataset [16], SNMP-MIB [2], and the KDD99 [23].

The Kitsune dataset was presented in [16]; a dataset of nine attacks that are generated on a video surveillance network and a Wi-Fi network with 3PCs and nine other IoT-related devices. The data related to these attacks were collected in separate CSV files. In this study, we are using the ARP poisoning attack dataset, a man-in-the-middle attack with a corresponding dataset of 116 features. The SNMP-MIB dataset was presented in [2], where the authors generated six types of DoS attacks and a Brute Force attack in a Simple Network Management Protocol (SNMP) environment and generated a realistic dataset of 34 features. The famous KDD99 dataset is of 41 features; it was detailed and analyzed in [23]. The three datasets are tabular and stored in CSV files.

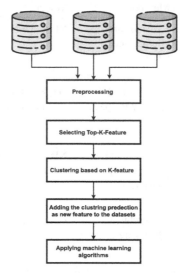

Fig. 1. The Methodology Applied through this Work

The methodology followed through this work is presented in Fig. 1. In the pre-processing step, the SNMP-MIB dataset was transformed from a multi-class dataset into a binary-class dataset. Every non-normal packet was considered anomalous, while normal packets were kept in the class labeled Normal. All categorical features in the three datasets were transformed into numerical features. In the second step, the top five features for every dataset were investigated using the SelectKBest and mutual-info-classif libraries of Python's sklearn. Then these features are used to select the most essential k features based on the entropy level between the features and the *class* column in every dataset.

In the third and fourth steps, every dataset was clustered using the k-means and DBSCAN methods. The two clustering methods were applied to every dataset based on the top two important features. Meanwhile, the cluster centroids were set to two. Then the clustering methods were applied to every dataset based on the top three important features, and cluster centroids were set to three. The same steps were repeated for the top four features and the top five features.

Two clustering methods on three datasets were applied four times every time with a different number of clusters, making them 2 * 3 * 4 = 24 clustering operations. For example, when the number of clusters is two, the k-means clustering was performed on the top 2, 3, 4 and 5 features in each dataset. The process repeats with density-based clustering (DBSCAN). Every time number of clusters to generate (k) was set to a number similar to the number of features that were used through the clustering process.

Then in the next step, we add the clustering prediction as a new feature to the datasets. Every clustering result generated from the 24 clustering operations was appended to the corresponding dataset independently; only one extra feature was appended at a time.

Then, the dataset with the appended new feature from the previous step was trained and tested with nine ML models. The models used in this study are KNN, SVM, Naive Bays, Linear regression, LDA, C4.5, XG-Boost, Random Forrest, and Ada. A 10-K fold statistical analysis was applied for every model with every dataset. The mean value for the ten folds was then measured.

Finally, the results generated from these models were compared for each dataset. The comparisons were also held between the cases when clustering was used before classification and when no clustering was applied.

4 Results and Discussion

This section shows the effects of cluster decomposing on the datasets mentioned above, where the cluster ID is added as a new feature.

Two hundred and forty three experiments were conducted (2 * 3 *4 * 9 + 3* 9 = 243). Each dataset has been clustered using the top two features and added the results to the dataset as a new feature. After that, the datasets were fed to nine ML models (Ada, C4.5, KNN, LDA, LR, NB, RF, SVM, and XGB). This process was repeated for the top three, four, and five features. In addition, the three datasets were also used without any clustering in the ML models to study the impact of cluster decomposing on the classification step.

The Five common classification evaluation metrics (accuracy, recall, precision, F1 measure, and AUC) have been used to study the impact of using this process. In addition to these metrics, MCC has also been used.

As illustrated in Figs. 2, 3, 4, 5, and 7, it can be noticed that when using DBSCAN for adding the cluster data as a new feature to the dataset, yields slightly better results in terms of accuracy, precision, recall, F1 measure, and MCC. However, this is not applied to AUC, as shown in Fig. 6 using the original dataset without using the clustering ID has the best result.

From Tables 2 and 7, to achieve the best accuracy and MCC for the ARP dataset, LDA should be used without any clustering technique. However, DB

SCAN clustering and XGB can enhance the precision, recall, and F1 measure. For Better AUC for the ARP dataset, KNN without using any clustering is recommended.

On the other hand, the KDD99 dataset, using DBSCAN clustering with the XGB model, yields better results for all used metrics except for the AUC, where the NB model without any clustering leads to better results, as shown in Tables 2, 3, 4, 5, 6, and 7.

For the SNMP dataset, using the ADA model can achieve better results than any other ML models in terms of accuracy, precision, recall, and f1-measure. In terms of MCC, using k-means cluster along with the random forest classifier is preferable. While the best AUC for SNMP-MIB dataset can be conducted using k-means with the C4.5 classifier.

It is worth mentioning that, the enhancement of the models in all cases did not exceed 1%. In other words, using the cluster decomposing technique does not alleviate the models' performance. In the case of intrusion detection datasets, the enhancement provides minimal improvement (Figs. 2, 3, 4, 5 and 7).

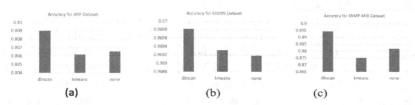

Fig. 2. Average of Accuracy

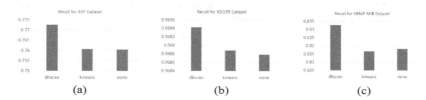

Fig. 3. Average of Recall

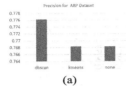
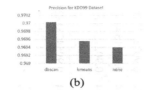
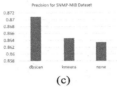

Fig. 4. Average of Precision

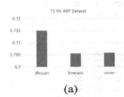
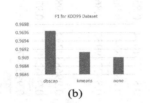
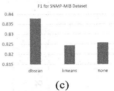

Fig. 5. Average of F1-measure

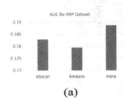
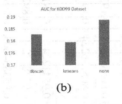
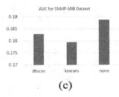

Fig. 6. Average of AUC

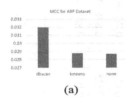
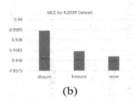
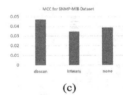

Fig. 7. Average of MCC

Table 2. The Average of Accuracy

Row Labels	Ada	C4.5	KNN	LDA	LR	NB	RF	SVM	XGB
ARPDF									
dbscan	0.8794465744	0.885499589	0.884419268	0.911464154	0.864769558	0.215070358	0.873761074	0.859552891	0.907337748
kmeans	0.88908241	0.867049834	0.874082277	0.914248744	0.865852891	0.195241183	0.882178741	0.865936224	0.902154414
none	0.89018241	0.867049834	0.874082277	0.91491541	0.865736224	0.196291038	0.882779931	0.865936224	0.902154414
NetDF									
dbscan	0.964919509	0.995167131	0.995315995	0.954112361	0.9541224	0.891910378	0.996834294	0.980579174	0.995425182
kmeans	0.964314102	0.994254046	0.994035737	0.954261257	0.953884218	0.89266472	0.99628847	0.979309017	0.994839647
none	0.964314102	0.994313665	0.994404565	0.954072698	0.953834603	0.891315022	0.996387704	0.979557049	0.994839647
SNMPDF									
dbscan	0.9251	0.866	0.9032	0.869505411	0.9129	0.88155	0.91325	0.9042	0.8727
kmeans	0.9256	0.8391	0.88935	0.848947094	0.88255	0.8605	0.89315	0.8853	0.8518
none	0.9256	0.8535	0.8948	0.863947094	0.8968	0.8604	0.8975	0.8914	0.8518

Table 3. The Average of Recall

	Ada	C4.5	KNN	LDA	LR	NB	RF	SVM	XGB
ARPDF									
dbscan	0.937001796	0.92765765	0.726719897	0.754079765	0.772317324	0.204162621	0.909025058	0.770963529	0.951569171
kmeans	0.942027928	0.930512662	0.657825086	0.75554497	0.745636364	0.1943894	0.938419652	0.732870035	0.948860112
none	0.942602841	0.930512662	0.657825086	0.755879058	0.746682632	0.194909413	0.934878356	0.732870035	0.948860112
NetDF									
dbscan	0.964597228	0.995119382	0.99528429	0.953231476	0.953038887	0.891593486	0.996742501	0.979978099	0.995253592
kmeans	0.964032526	0.99419971	0.993983644	0.953380742	0.952804331	0.892094668	0.996204633	0.978583503	0.994660281
none	0.964032526	0.994276747	0.993985943	0.953189967	0.952751228	0.890992005	0.996319351	0.978877801	0.994660281
SNMPDF									
dbscan	0.8836875	0.80059375	0.84074375	0.769708955	0.83735	0.8383	0.857125	0.84351875	0.82414375
kmeans	0.884	0.776875	0.82734375	0.741636047	0.824525	0.82775	0.8353	0.823175	0.809925
none	0.884	0.7850625	0.832225	0.729298547	0.834925	0.8277	0.82903125	0.831875	0.809925

Table 4. The Average of Precision

Row Labels	Ada	C4.5	KNN	LDA	LR	NB	RF	SVM	XGB
ARPDF									
dbscan	0.90723787	0.895469713	0.721553458	0.711055273	0.720609963	0.505535917	0.8822399	0.729682532	0.912887237
kmeans	0.908260564	0.906230908	0.65844117	0.711538909	0.707270624	0.505796665	0.907959586	0.699904881	0.909391908
none	0.908438477	0.906230908	0.65844117	0.711541333	0.707854846	0.505786399	0.907341873	0.699904881	0.909391908
NetDF									
dbscan	0.964916671	0.995170738	0.995306503	0.954564732	0.954856508	0.891927536	0.996898636	0.981058301	0.995560266
kmeans	0.964275708	0.994256334	0.994031908	0.954715572	0.954606351	0.892875815	0.996339863	0.979935268	0.994977947
none	0.964275708	0.994300878	0.99404918	0.954526168	0.954561385	0.891372115	0.996422539	0.980115626	0.994977947
SNMPDF									
dbscan	0.939042135	0.846939322	0.868981942	0.812715422	0.861543275	0.875125628	0.897400609	0.871550524	0.864343081
kmeans	0.942372881	0.837261215	0.863865622	0.791275912	0.86602974	0.875000313	0.886944957	0.859874618	0.858722233
none	0.942372881	0.83833346	0.866111121	0.774154589	0.870219023	0.875	0.877927497	0.867922794	0.858722233

Table 5. The Average of F1 measure

Row Labels	Ada	C4.5	KNN	LDA	LR	NB	RF	SVM	XGB
ARPDF									
dbscan	0.865053894	0.856418109	0.686904984	0.691898006	0.716167658	0.177670536	0.834838561	0.721099198	0.885954529
kmeans	0.870890069	0.857659512	0.618686332	0.693835713	0.702404554	0.167557201	0.865512014	0.693340382	0.880081278
none	0.871590784	0.857659512	0.618686332	0.694024597	0.703087545	0.168310624	0.8657215	0.693329155	0.880081278
NetDF									
dbscan	0.964730016	0.995142552	0.995293324	0.953810063	0.95379933	0.89137253	0.996817607	0.980456446	0.995400361
kmeans	0.964122206	0.994225207	0.99400474	0.953960003	0.953559694	0.8920077498	0.996268992	0.97917354	0.994811189
none	0.964122206	0.994285416	0.994014362	0.953770001	0.953509515	0.890771596	0.996368796	0.979424405	0.994811189
SNMPDF									
dbscan	0.895764717	0.806685086	0.845182986	0.778136355	0.839909841	0.832173278	0.869693135	0.848583691	0.824226357
kmeans	0.896317988	0.78619703	0.836124536	0.749706324	0.835126422	0.825782412	0.848502855	0.832261936	0.810069547
none	0.896317988	0.79183952	0.840276372	0.737665183	0.844230359	0.825697025	0.844531245	0.841979662	0.810069547

Table 6. The Average of AUC

	Ada	C4.5	KNN	LDA	LR	NB	RF	SVM	XGB
ARPDF									
dbscan	0.211696887	0.298961751	0.351650518	0.128029547	0.051802466	0.235194849	0.146906751	0.116061564	0.105258388
kmeans	0.186592833	0.347436692	0.371541239	0.129869186	0.042956312	0.221580174	0.130096416	0.110188702	0.075397761
none	0.186592833	0.347436692	0.371541239	0.129889278	0.040997052	0.221580174	0.216267425	0.10865292	0.075397761
NetDF									
dbscan	0.005461864	0.004871938	0.00471571	0.011509308	0.010065407	0.045023955	0.00028647	0.002549557	0.000297811
kmeans	0.005436357	0.005764004	0.006016357	0.011484941	0.010163032	0.044398889	0.000370386	0.002890929	0.000295684
none	0.005436357	0.005686816	0.006014057	0.011498009	0.010159709	0.045563983	0.000316388	0.002862661	0.000295684
SNMPDF									
dbscan	0.0135	0.4953125	0.2278125	0.0881625	0.0764625	0.0456	0.087140625	0.0975375	0.2211875
kmeans	0.018	0.63375	0.3078125	0.14934375	0.16625	0.04454375	0.078009375	0.2757875	0.25645
none	0.018	0.584375	0.27875	0.125225	0.137225	0.04565	0.135575	0.1612	0.25645

Table 7. The Average of MCC

Row Labels	Ada	C4.5	KNN	LDA	LR	NB	RF	SVM	XGB
ARPDF									
dbscan	0.032545872	0.035662797	0.03248216	0.052282699	0.022082779	0.022071522	0.030398056	0.009139305	0.051374036
kmeans	0.03709691	0.027463119	0.029372078	0.054127243	0.014005874	0.024132941	0.034542413	-0.004493073	0.042826847
none	0.037760635	0.027463119	0.029372078	0.054140296	0.015661156	0.024060022	0.031716328	-0.004493073	0.042826847
NetDF									
dbscan	0.929513567	0.990290102	0.990590765	0.907794953	0.907893063	0.783520344	0.993641101	0.96103554	0.990813793
kmeans	0.928307858	0.988456022	0.9888015507	0.908095045	0.907408379	0.784968758	0.992544458	0.958517621	0.989638147
none	0.928307858	0.988577602	0.988035076	0.90771488	0.9073110291	0.782363463	0.992741864	0.958992435	0.989638147
SNMPDF									
dbscan	0.046156883	-0.00283471	0.04545506	0.061628335	0.057067925	0.061493146	0.067457243	0.051455116	0.035502054
kmeans	0.04871223	-0.02607149	0.032600407	0.040392992	0.036689587	0.061237616	0.070867538	0.025051845	0.021802791
none	0.04871223	-0.017526949	0.03776022	0.059486531	0.045966161	0.061237244	0.050201432	0.04180278	0.021802791

5 Conclusion

An empirical study has been conducted in the domain of Intrusion Detection. Clustering algorithms have highlighted the similarities between datasets instances and grouped them accordingly. The resulting cluster id has been added as a feature in the original dataset, which is further used to detect intrusion using several classifiers. Several combinations have been tested in this study, involving the clustering method, the number of clusters, and the classifiers. Several evaluation metrics have been used to evaluate the performance of the classification part. DBSCAN has yielded slight enhancement in all metrics except for the AUC, which was better without applying the clustering phase.

As future work, it is intended to test the aforementioned combinations on multi-labeled datasets and check whether the clustering step would affect the classification performance in this case. Also, different types of data augmentation should be tested.

References

1. Ahmad, Z., Shahid Khan, A., Wai Shiang, C., Abdullah, J., Ahmad, F.: Network intrusion detection system: a systematic study of machine learning and deep learning approaches. Trans. Emerg. Telecommun. Technol. **32**(1), e4150 (2021)
2. Al-Kasassbeh, M., Al-Naymat, G., Al-Hawari, E.: Towards generating realistic SNMP-MIB dataset for network anomaly detection. Int. J. Comput. Sci. Inf. Secur. **14**(9), 1162 (2016)
3. Alapati, Y.K., Sindhu, K.: Combining clustering with classification: a technique to improve classification accuracy. Lung Cancer **32**(57), 3 (2016)
4. Alshaibanee, A.F., AlJanabi, K.B.: A proposed class labeling approach: From unsupervised to supervised learning. In: 2022 Iraqi International Conference on Communication and Information Technologies (IICCIT), pp. 1–6. IEEE (2022)
5. Arora, P., Varshney, S., et al.: Analysis of k-means and k-medoids algorithm for big data. Procedia Comput. Sci. **78**, 507–512 (2016)
6. Aziz, A.S.A., Sanaa, E., Hassanien, A.E.: Comparison of classification techniques applied for network intrusion detection and classification. J. Appl. Log. **24**, 109–118 (2017)
7. Barmpas, P., et al.: A hybrid machine learning framework for enhancing the prediction power in large scale population studies: The athlos project. medRxiv, pp. 2021–01 (2021)
8. Bhattacharjee, P., Mitra, P.: A survey of density based clustering algorithms. Front. Comput. Sci. **15**, 1–27 (2021)
9. Chang, C.C., Chen, S.H.: Developing a novel machine learning-based classification scheme for predicting SPCs in breast cancer survivors. Frontiers **10**, 848 (2019)
10. Gamage, S., Samarabandu, J.: Deep learning methods in network intrusion detection: a survey and an objective comparison. J. Netw. Comput. Appl. **169**, 102767 (2020)
11. Jader, R., Aminifar, S., et al.: Predictive model for diagnosis of gestational diabetes in the kurdistan region by a combination of clustering and classification algorithms: an ensemble approach. Appl. Comput. Intell. Soft Comput. **2022** (2022)
12. Khraisat, A., Gondal, I., Vamplew, P., Kamruzzaman, J.: Survey of intrusion detection systems: techniques, datasets and challenges. Cybersecurity **2**(1), 1–22 (2019)
13. Kuttattu, A.S., Gokul, G., Prasad, H., Murali, J., Nair, L.S.: Analysing the learning style of an individual and suggesting field of study using machine learning techniques. In: 2019 International Conference on Communication and Electronics Systems (ICCES), pp. 1671–1675. IEEE (2019)

14. Manocha, S., Bansal, V., Kaushal, I., Bhat, D.A.: Efficient power theft detection using smart meter data in advanced metering infrastructure. In: Proceedings of the International Conference on Intelligent Computing and Control Systems (ICICCS 2020). IEEE (2020)
15. Mathivanan, N.M.N., Ghani, N.A.M., Janor, R.M.: Improving classification accuracy using clustering technique. Bull. Electr. Eng. Inform. 7(3), 465–470 (2018)
16. Mirsky, Y., Doitshman, T., Elovici, Y., Shabtai, A.: Kitsune: an ensemble of autoencoders for online network intrusion detection. arXiv preprint arXiv:1802.09089 (2018)
17. Muntean, M., Militaru, F.D.: Design science research framework for performance analysis using machine learning techniques. Electronics 11 (2022)
18. Piernik, M., Morzy, T.: A study on using data clustering for feature extraction to improve the quality of classification. Knowl. Inf. Syst. 63, 1771–1805 (2021)
19. Salih, A.A., Abdulazeez, A.M.: Evaluation of classification algorithms for intrusion detection system: a review. J. Soft Comput. Data Mining 2(1), 31–40 (2021)
20. Saranya, T., Sridevi, S., Deisy, C., Chung, T.D., Khan, M.A.: Performance analysis of machine learning algorithms in intrusion detection system: a review. Procedia Comput. Sci. 171, 1251–1260 (2020)
21. Shalaby, M., Belal, N.A., Omar, Y.: Data clustering improves siamese neural networks classification of Parkinson's disease. Complexity 2021, 1–9 (2021)
22. Talaei Khoei, T., Kaabouch, N.: A comparative analysis of supervised and unsupervised models for detecting attacks on the intrusion detection systems. Information 14(2), 103 (2023)
23. Tavallaee, M., Bagheri, E., Lu, W., Ghorbani, A.A.: A detailed analysis of the KDD cup 99 data set. In: 2009 IEEE Symposium on Computational Intelligence for Security and Defense Applications. pp. 1–6. IEEE (2009)
24. Yang, Z., et al.: A systematic literature review of methods and datasets for anomaly-based network intrusion detection. Comput. Secur. 102675 (2022)

Personalized Movie Recommendation Prediction Using Reinforcement Learning

Abderaouf Bahi[✉], Ibtissem Gasmi, and Sassi Bentrad

Department of Computer Science, Chadli Bendjedid El Tarf University, El Tarf, Algeria
{a.bahi,gasmi-ibtissem,bentrad-sassi}@univ-eltarf.dz

Abstract. Current recommendation systems do not always meet users' needs because they are mainly based on demographic data and consumption histories. However, reinforcement learning makes it possible to continuously improve the accuracy of these recommendations taking into account the real-time interactions of the user with the system. This paper deals with the application of a model based on reinforcement learning for recommending movies to users on an online movie platform with a dataset of movie ratings (Movie Lens 100K). The problem is how to use this method to suggest relevant books to users based on their preferences and reading histories to get better results than other methods. The evaluation metrics used in this study are precision and recall. The main results show that the proposed reinforcement learning algorithm outperformed the baseline methods in terms of precision and recall. Specifically, the algorithm achieved a precision of 0.67 and a recall of 0.51 which represents a significant improvement compared to the baseline methods. In this case, reinforcement learning in the recommendation system is found to be more effective in improving the recommendations with an increase in recommendation accuracy of up to 7% in some cases.

Keywords: Reinforcement Learning · Recommendation System · Data Analysis · Prediction

1 Introduction

The personalized movie recommendation has become a crucial issue for online movies applications [1, 2]. It is well known that users are often overwhelmed by the vast number of films available and need help finding those that best suit their preferences. Automated recommendation systems (RS) are used to address this effect, by suggesting movies that match the tastes and preferences of each user. The research questions of the proposed approach are focused on investigating how reinforcement learning can improve the effectiveness of recommendation systems. The study aims to explore the potential of reinforcement learning to enhance user satisfaction by addressing the limitations and challenges of existing systems.

J. Abawajy et al. (Eds.): ICICCT 2023, CCIS 1841, pp. 46–56, 2023.
https://doi.org/10.1007/978-3-031-43838-7_4

Existing recommendations systems face limitations and challenges such as low accuracy and scalability, In this work, a reinforcement learning (RL) based approach to predict personalized movie recommendations is used, which takes into account the long-term reward of recommendations instead of just focusing on the immediate accuracy of the predictions, leading to better user satisfaction and engagement with the recommendation system. The contribution of the proposed approach is to offer a novel and effective method for recommendation systems by leveraging the power of reinforcement learning. By providing a more personalized and accurate set of recommendations, the approach can enhance user satisfaction and engagement.

The database [3] used contains user ratings as well as information about the movies themselves, we analyzed and visualized it to understand its characteristics that influence the ratings given by users and then applied RL models to predict personalized movie recommendations for each user, taking into account their preferences and viewing history. After that, the considerate approach was evaluated using a recommendation accuracy measure, and also compared this approach to other recommendation methods, such as matrix factorization and collaborative filtering.

The results show that the RL-based approach is capable of predicting personalized movie recommendations with high accuracy. It was found that this approach is more effective than the compared recommendation methods [4]. Overall, the approaches considered in this work are vividly presented, their strengths and weaknesses, and future directions for improving the accuracy and relevance of personalized movie recommendations.

The rest of this work is structured as follows: in Sect. 2, an overview of the related work in the field of recommendation systems with a focus on RL-based recommendation systems is provided. The advantages and limitations of existing approaches are highlighted. The novelty of the proposed approach is also discussed. In Sect. 3, the proposed methodology including a film recommendation based on the RL method is carefully described. At first, an overview of the dataset used for the experiments is presented. Then, the preprocessing steps applied to the data to prepare it for the RL agent training are described. Moreover, the details of the RL algorithm and evaluation metrics used to assess the performance of the system are presented. Section 4 presents the data sources and details on the procedure used to experiment with the RL-based recommender system and the analysis of its effectiveness. Section 5 presents the results of the experiments and evaluates the performance. Finally, Sect. 6 presents conclusions and perspectives.

2   Related Work

Recommendation systems (RS) aim to provide personalized recommendations to users based on their preferences and interests [5]. They have become increasingly important in the era of big data and online services, where the amount of available information is overwhelming and users need assistance to find relevant content. In the literature, we distinguish mainly from RS of which content-based collaborative filtering (CF) [6]. While content-based RS use item characteristics to recommend similar items to users, collaborative filtering recommender systems use user preferences to recommend items that similar users like, A context-aware recommender system (CARS) [7] applies sensing and analysis of user context to provide personalized services.

A hybrid approach combining content-based and collaborative filtering techniques has also been proposed [2]. However, traditional recommendation systems have been successful in many applications, they have some limitations. Traditional recommendation approaches, such as matrix factorization and collaborative filtering methods, have significant limitations in terms of accuracy and ability to process large-scale data. Indeed, these approaches tend to suffer from problems such as difficulty in dealing with missing data, popularity bias and overfitting. Moreover, they are not always effective in capturing complex user preferences, such as higher-order interactions and contextual preferences.

Sharma et al. [8] provided a comprehensive overview of research paper recommender systems. They outlined research paper recommendation and highlighted its significance in academia. The paper examines various approaches, including content-based, collaborative filtering, and hybrid recommendation techniques, and discusses their strengths and weaknesses. Moreover, the authors reviewed various evaluation metrics and datasets used to assess the performance of research paper recommender systems. They also explored the role of deep learning techniques in enhancing the accuracy of research paper recommendation.

In their paper [9], Jayalakshmi et al. highlighted the importance of recommender systems and the difficulties encountered in the movie domain. They examined various data types employed in movie recommendation, such as user ratings, item features, and contextual information. Additionally, they explored how deep learning techniques can enhance the accuracy of movie recommendation. The paper concludes by suggesting potential future directions for movie recommender systems, such as incorporating social networks, utilizing explainable AI, and developing personalized recommendation systems.

2.1 Reinforcement Learning for Recommendation

Reinforcement Learning has recently emerged as a promising approach to address the limitations of traditional recommendation systems [10, 11]. RL-based recommendation systems can learn to recommend items by interacting with users and receiving feedback in the form of rewards. The rewards can be defined based on the user's feedback, such as the rating or click-through rate, or based on the system's objectives, such as diversity or novelty. RL-based recommendation systems have several advantages over traditional recommendation systems. First, they can handle the cold start problem by exploring the user's preferences through interaction. Second, they can handle the sparsity problem by leveraging the feedback received from the user to improve the recommendations. Third, they can optimize different objectives, such as diversity or novelty, based on the reward function [12]. Several RL-based recommendation systems have been proposed in the literature. For example, in [12] the authors propose a deep RL approach to learn to recommend news articles based on user clicks. They used a dueling network architecture to estimate the Q-values of the recommendations and used an epsilon-greedy exploration strategy and in [13] a deep RL approach to learn to recommend songs based on the user's listening history is proposed. They used a policy gradient algorithm to optimize the recommendations and included a diversity term in the reward function to encourage diverse recommendations.

Overall, the use of RL in recommender systems is a promising approach to address the limitations of traditional approaches. However, this requires a thorough understanding of how RL can be effectively applied to recommender systems, as well as the potential challenges and limitations of this approach.

2.2 Novelty of Our Approach

Regarding our contribution in this work is the use of a simple representation of the items and the user's preferences, is beneficial because it reduces the complexity of the recommendation process. By representing items and preferences in a simplified form, the algorithm can efficiently process a large amount of data and make recommendations based on user preferences. This also allows for easier interpretation and understanding of the recommendations, which can improve user satisfaction and trust in the system. Additionally, simpler representations can make the recommendation algorithm more robust to changes in user behavior and preferences over time. A similar study [14] also used a representation of user preferences and object characteristics for the recommendation. However, their method did not take into account the user's past interactions with the recommended objects. In contrast, the approach proposed in this paper uses reinforcement to dynamically learn user preferences over time, allowing for more accurate and personalized recommendation.

After that, appropriately representation articles based on genre, include a bias term in the user's input vector to capture their general preference for movies, and use a dual network architecture to determine the Q-value increase of recommendations for improving the stability and convergence of the RL algorithm. Moreover, the proposed approach is swallowed by real-world movie rating datasets [15–18]. These references discuss how RL is used by Netflix and Amazon Prime Video to optimize and personalize content recommendations. This allows them to demonstrate real-world effectiveness and compare it with existing recommendation systems [19, 20].

To this end, The Dueling Network architecture was proposed to improve the stability and convergence of the reinforcement learning algorithm. It separates the value and advantage functions of the Q-function, which estimates the expected future reward for each action taken in a given state. The value function represents the value of being in a particular state regardless of the action taken, while the advantage function measures how much better each action is than the average action in that state

3 Methodology

This study aims to answer if a reinforcement learning approach that incorporates dueling network architecture and a simple representation of user preferences and item features can improve the accuracy and stability of movie recommendation systems compared to traditional collaborative filtering and content-bases methods.

To test and validate the proposed recommendation system, the movie's dataset is considered [3].The latter dataset contains movie ratings from popular online movie recommendation services (see Tables 1 and 2). We use the training set to train the RL agent and the test set to evaluate the performance of the system (Tables 4 and 5).

Table 1. Movie database

Movield	Title	Genres
1	Toy Story	Adventure, Children, Fantasy
2	Jumanji	Comedy, Romance
3	Grumpier Old Men	Comedy, Drama, Romance
4	Waiting to Exhale	Adventure, Drama, Romance

Table 2. Rating database

Movield	Title	Genres
1	16	4.0
1	24	1.5
1	47	4.0
1	50	3.0

3.1 Data Cleaning and Visualization

Before training the RL agent, some cleaning procedures are applied to the data. First, we normalized the ratings by subtracting each user's average rating from the rating. This helped reduce the impact of individual user biases on recommendations. The ratings are then converted into binary feedback. A rating of 4 or higher is considered positive feedback, and a rating of less than 4 is considered negative feedback. The movie genre information is encoded using one-hot encoding. This allowed us to include genre information in the RL agent's country display for better recommendations.

Finally, in this study the threshold of 4 was chosen as the threshold value to determine positive reviews because it corresponds to the average rating given to films by users in the Netflix database [15, 16]. This value was considered a reasonable starting point for defining the boundary between positive and negative reviews.

In contrast, the threshold value of less than 4 for negative reviews was chosen arbitrarily to have a balance between positive and negative reviews in the data set. It is important to note that this threshold can be adjusted depending on the specific study objectives and user preferences.

Other thresholds have been considered and tested in the literature depending on the specific objectives of each study and the characteristics of the data used.

Figure 1 displays a bar chart showing the number of views of the most popular movies in the MovieLens 100K dataset. The horizontal axis represents the movie ID, while the vertical axis represents the number of views. Each bar on the chart corresponds to a single movie, and its height indicates the number of views received. The chart reveals that some movies, such as those with ID 296, 356, and 318, have received a significantly larger number of views than others. Additionally, the chart illustrates the relative popularity of different movies, with the most viewed movies having considerably higher bars than the least viewed ones. This helps in identifying the most and least viewed movies in the dataset and understanding how user behavior and preferences differ across various movies.

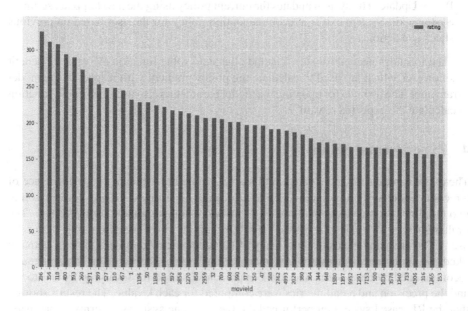

Fig. 1. Most Popular movies.

3.2 RL Algorithm

In this work, A Q-learning algorithm with a dueling network architecture is used to train RL agents [21]. The Dueling Network architecture separates the estimation of state and profit values. This has been shown to improve algorithm stability. The input to the dueling network consisted of the user's feedback history, represented as a set of binary his feedback values, and one-hot-encoded movie genre information [22]. We also included a bias term in the user input vector to capture overall movie preferences. Importantly, the output of the dual network was the Q-value for each possible action, where the action represented a movie recommendation. Agents chose the action with the highest Q value and were rewarded based on user feedback.

To facilitate exploration, we used an epsilon-greedy search strategy, in which agents choose random actions with probability epsilon and optimal actions with probability 1-epsilon [17]. Here is a description of each step of the reinforcement learning algorithm proposed in the study:

- Initialization: The system initializes the model parameters;
- Action Selection: The system selects an action from the current policy based on the current state and template parameters;
- Action execution: The system executes the selected action and observes the reward and the next state;
- Model Parameter Update: The system updates the model parameters using the gradient descent method to maximize the expected reward;
- Policy Update: The system updates the current policy using the new template settings;
- System exit: System exit is a recommendation policy for the user based on previous states and actions.

The formulas needed for step 4 include the state value function (V) and the benefit function (A), which are used to calculate the policy gradient with respect to the model parameters. The formula for updating the model parameters also uses the benefit function to calculate the expected reward.

4 Experimental Setup

The experimental study conducted in this article aimed to evaluate the performance of a recommender system based on deep reinforcement learning (DRL) in comparison to two baseline systems: a content-based system and a collaborative filtering system. The evaluation was done using the Movie Lens100k dataset, which contains movie ratings by users. The precision measures how many of the recommended movies the user actually liked, while the recall measures how many of the movies the user liked were actually recommended by the system. The top-k approach was used for k varying from 1 to 10, and the precision and recall metrics were computed for each k value. The results showed that the RL-based system outperformed the two baseline systems in terms of precision and recall, especially for larger k values. However, the performance of the RL-based system degraded as the number of recommended movies increased, which is a common problem in recommender systems.

4.1 Dataset

The dataset consists of his 100,000 ratings on a scale of 1 to 5 for his 1682 movies from 943 users [3], and also includes information about the genre of the movie. We used this to enhance the feature representation of the RL algorithm by splitting the dataset into training sets.

In order to prepare the data for RL-based recommendation, we performed several preprocessing steps. First, the removal of movies with less than 50 ratings, as it is unlikely that these movies are popular with users, and also the removal of users who had rated less than 10 movies, as it is unlikely that these users provided enough information about their preferences. Then, the remaining data are divided into training and test sets in an 80:20 split. The training set was used to train the Q function, while the test set was used to evaluate the performance of the LR-based recommender system. Furthermore, to encode the movie information in a way that can be used by the Q function, each movie is presented as a vector of binary features indicating the genre of the movie. As a result, a total of 19 binary features, one for each genre, were considered. For example, if a movie belongs to the drama and romance genres, the corresponding binary vector is [1, 0, 0, 0, 0, 0, 0, 1, 0, 0, 0, 0, 0, 0, 0, 0, 0, 0, 0]. And, to encode the user's viewing history, the history is represented as a vector of movie ratings, with each rating normalized to the interval [0, 1] while including a bias term in the user's input vector to capture the user's overall preference for movies.

5   Results and Discussion

This section presents the results obtained and the performance evaluation of the proposed system. It consists of movie recommendations based on RL and comparisons with other studies in the field [10, 11, 18].

5.1   Evaluation Metrics

Table 3 shows the precision and recall results for the proposed system and the two baseline systems, using the top-k approach with k varying from 1 to 10. Our RL-based system achieved the highest precision and recall for all values of k, indicating that it outperformed the baseline systems. The precision and recall values for our system were highest for $k = 1$, indicating that our system was most accurate in recommending a single movie to the user. As k increased, the precision and recall values decreased, indicating that our system struggled to recommend a larger number of movies.

Table 3. Precision calculated

k	RL-Bases System	Content-bases System	Collaborative filtering system
1	0.89	0.68	0.72
2	0.72	0.61	0.64
3	0.66	0.43	0.57
4	0.58	0.40	0.51
5	0.54	0.34	0.47
6	0.43	0.23	0.42
7	0.41	0.16	0.31
8	0.38	0.13	0.24
9	0.34	0.06	0.16
10	0.25	0.02	0.12

Table 4. Recall calculated

k	RL-Bases System	Content-bases System	Collaborative filtering system
1	0.52	0.28	0.31
2	0.38	0.21	0.24
3	0.31	0.17	0.20
4	0.26	0.14	0.17
5	0.23	0.12	0.15
6	0.20	0.11	0.13
7	0.18	0.10	0.12
8	0.16	0.09	0.11
9	0.15	0.08	0.10
10	0.14	0.07	0.09

5.2 Analysis of Recommendations

As part of the analysis for this work, a case study of 10 users in the test set was conducted. So, for each user, we looked at the top 5 recommended movies in the dataset and their ratings, as well as how they compare to the actual movies that the user rated highly. We found that the proposed system can better recommend movies that the user has not previously rated but are similar in genre and style to movies the user has previously enjoyed. For example, if a user rated an action movie highly, our system recommended his sci-fi movie that he had never seen before but had a similar style and tone to an action movie he liked.. In contrast, content-based systems recommended dramas that belonged to the same genre as action movies, but did not do well for users. In addition, it was difficult for the recommendation system proposed in this research to recommend movies that do not match the user's usual tastes. For example, a user rated his comedy Romantic, but the system had a hard time recommending a horror movie the user had never seen, even with positive reviews.

Table 5. Example of recommended movies for a random user

Film	Rating
Star Wars	4.9
The Shawshank Redemption	4.8
The Godfather	4.8
The Dark Knight	4.7
The Godfather: Part II	4.6

5.3 Limitations and Future Work

However, the proposed RL-based movie recommendation system has some limitations, such as the performance of the system degrades with the number of movies recommended, There are several reasons for this, one reason is that the user's attention span and interest may decrease as more recommendations are presented, making it harder for them to choose [23]. Additionally, as the number of recommendations increases, the diversity of the recommendations may decrease, leading to a less personalized experience for the user. Finally, there may be a higher chance of including irrelevant or disliked movies in the larger set of recommendations, leading to a lower overall satisfaction for the user. In addition to addressing the current limitations of the system, some more concrete suggestions for future work could involve algorithmic or methodological improvements. For example, one possible direction could be to explore the use of deep reinforcement learning techniques to better capture the complex interactions between users, items, and the environment. Another possibility would be to incorporate additional sources of information [24], such as contextual features or social network data, to further enhance the recommendation accuracy. Finally, it may also be worth considering alternative evaluation metrics that go beyond traditional accuracy measures, such as diversity or comprehensive assessment of the system's performance.

6   Conclusions and Perspective

In this work, a RL-based movie recommendation system that can provide users with personalized movie recommendations based on their previous movie ratings is proposed. As for its validation, it is shown to outperform the two reference systems of his compared in this study, especially the content-based system and the collaborative filtering system, in terms of accuracy and recall. It can recommend movies that the user has never seen but are similar in genre and style to movies the user has already seen. Unfortunately, it's difficult to recommend movies that don't match the user's usual tastes.

However, future work will target specific indexed bonds that will undoubtedly contribute to the development of recommendation systems. One possible direction for future work is to explore using deep reinforcement learning algorithms to improve performance. Another direction is to integrate more data sources into the system, such as recommendations.

References

1. Chen, M., Gao, Y., Liu, Y.: A survey of collaborative filtering-based recommender systems. IEEE Trans. Ind. Inf. **16**(4), 2233–2249 (2020)
2. Cheng, Y., Lu, X., Xu, J.: A hybrid recommendation method for personalized news articles. Neurocomputing **440**, 1–11 (2021)
3. Book Crossing. (n.d.). How to BookCross (2023). https://www.bookcrossing.com/howto. Accessed 25 2023
4. Chen, L., Wang, X., Zhang, B.: A survey on deep learning for recommender systems. Neurocomputing **396**, 411–427 (2020)

5. Guo, H., Zhang, Y., Fan, X., Jin, R.: Recurrent neural networks with auxiliary information for personalized product search. Inf. Retrieval J. **23**(3), 244–261 (2020)
6. Gasmi, I., Anguel, F., Seridi-Bouchelaghem, H., Azizi, N.: Context-aware based evolutionary collaborative filtering algorithm. In: Chikhi, S., Amine, A., Chaoui, A., Saidouni, D.E., Kholladi, M.K. (eds.) MISC 2020. LNNS, vol. 156, pp. 217–232. Springer, Cham (2021). https://doi.org/10.1007/978-3-030-58861-8_16
7. Gasmi, I., Azizi, M.W., Seridi-Bouchlaghem, H., Azizi, N., Belhaouari, S.M.: Enhanced context-aware recommendation using topic modeling and particle swarm optimization. J. Intell. Fuzzy Syst. **40**(6), 12227–12242 (2021)
8. Sharma, R., Gopalani, D., Meena, Y.: An anatomization of research paper recommender system: overview, approaches and challenges. Eng. Appl. Artif. Intell. **118**, 105641 (2023)
9. Jayalakshmi, S., Ganesh, N., Cep, R., Senthil Murugan, J.: Movie ˇrecommender systems: concepts, methods, challenges, and future directions. Sensors **22**, 4904, (2022)
10. Pan, R., Chen, L.: Efficient neural collaborative filtering with binary quantization. IEEE Trans. Knowl. Data Eng. **33**(4), 1602–1614 (2020)
11. Li, J., Li, Y., Zhang, X., Li, Y., Liu, Y.: Multi-task learning for personalized product search. ACM Trans. Inf. Syst. (TOIS) **38**(1), 1–24 (2020)
12. Mnih, et al.: Human-level control through deep reinforcement learning. Nature **518**, 529–533 (2015)
13. He, K., Zhang, X., Ren, S., Sun, J.: Deep residual learning for image recognition. In: Proceedings of the IEEE Conference on Computer Vision and Pattern Recognition, pp 770- 778 (2016).https://doi.org/10.1109/CVPR.2016.90
14. Koren, Y., Bell, R.: Advances in collaborative filtering. In: Ricci, F., Rokach, L., Shapira, B., Kantor, P.B. (eds.) Recommender Systems Handbook, pp. 145–186. Springer US (2007)
15. Ma, K., et al.: Using Machine Learning to Improve Streaming Quality at Netflix (2015)
16. Khaku, S., et al.: Improving Netflix's Streaming Infrastructure with Machine Learning (2018)
17. Jain, Y.K., et al.: Optimizing Video Streaming Using Machine Learning at Amazon Prime Video (2018)
18. Large Scale Video Streaming with Machine Learning at Amazon Prime Video (2021)
19. Huang, Y., Guo, X., Cao, X., Hu, X.: Improved collaborative filtering algorithm based on tag semantic analysis. J. Ambient. Intell. Humaniz. Comput. **11**(12), 5327–5335 (2020)
20. Li, X., Gao, Y., Liu, Y., Wu, L.: Learning from reviews: sentiment-aware neural recommendation. IEEE Trans. Neural Netw. Learn. Syst. **31**(9), 3445–3457 (2020)
21. Wu, X., Zhang, Y., Wang, C.: A survey on deep learning for recommendation. Artif. Intell. Rev. **53**(2), 1647–1672 (2020)
22. Yu, D., Zhang, W.: A hybrid model for top-N recommendation with implicit feedback. Inf. Process. Manage. **57**(5), 102275 (2020)
23. Yang, Z., Xue, H., Xu, W., Liu, Y.: Easing the use of unlabeled data in deep neural networks for personalized recommendation. IEEE Trans. Knowl. Data Eng. **32**(5), 980–994 (2020)
24. Xiang, H., Cheng, X., Song, J., Hu, W., Chen, F.: Incorporating diversity into deep neural networks for recommender systems. Neural Netw. **124**, 239–250 (2020)

An Ensemble and Deep Neural Network Based Approaches for Automated Sentiment Analysis

Riya$^{(\boxtimes)}$ ⓘ, Sonali Rai ⓘ, Rupal ⓘ, Ritu Rani ⓘ, Vandana Niranjan ⓘ, and Arun Sharma ⓘ

Indira Gandhi Delhi Technical University for Women, New Delhi, India
riya010btece19@igdtuw.ac.in

Abstract. The rise of technology in social media has resulted in an enormous quantity of textual ambiguity. Sentiment analysis provides a crux of subjective opinions stored in a large amount of text such that the data gets segregated into positive and negative. In this research two datasets are used i.e., Amazon Reviews and IMDb wherein, first, implemented machine learning models such as Naive Bayes, XGBoost, etc. for the sentiment analysis out of which Linear SVC performed the best for the IMDb dataset and Amazon Reviews dataset. Furthermore, implemented a deep learning model i.e., Bi-LSTM that outperformed machine learning models for both datasets. Next, implemented BERT, a pre-trained language model, showed better results than Bi-LSTM for both datasets. Lastly, proposed a hybrid CNN-LSTM model, wherein, for feature extraction CNN is used, while LSTM is used for classification. The proposed hybrid model has given the best ROC score for both datasets among all the models used in this paper.

Keywords: Sentiment Analysis · Machine Learning · Deep Learning · Natural Language Processing · Hybrid Model · Recurrent Neural Network · Convolutional Neural Network · Long Short-Term Memory · Bidirectional Encoder Representations from Transformers · Linear SVC · Ensemble Model · Random Forest

1 Introduction

The ongoing generation, popularly known as Gen Z has taken over the concept of the Internet like no other generation. As Gen Z is accustomed to using social media, they are also extremely straightforward about their thoughts and use social media as a tool to express their opinion be it about elections, a product, or a movie. From buying a product to watching a movie, nowadays, people prefer taking a decision based on reviews to save time. In order, to understand the choices of their customers well, companies have given the option to get a review from their customers. However, organizing such a huge amount of data is a challenging task. So, to ease this issue of whether the product serves its purpose or not, sentiment analysis is an advantageous choice. What role does sentiment analysis play? What is the need to understand the sentiments behind any text? What difference does it make if we know the intentions trailing the texts? To answer this, take the following examples.

J. Abawajy et al. (Eds.): ICICCT 2023, CCIS 1841, pp. 57–73, 2023.
https://doi.org/10.1007/978-3-031-43838-7_5

- When someone wants to buy a product, they look up the reviews of the product before making any decision.
- Similarly, before watching any movie people like to check its IMDb rating to save themselves from wasting their time.

It all started with the invention of machine translation which was founded in the time of the second world war in 1940s to convert Russian language to English. The study on fundamental and prospective themes such as word sense disambiguation and statistically colored NLP, as well as lexicon work, were given a study focus. Furthermore, the pursuit of NLP was accompanied by other crucial concerns such as statistical language processing, information retrieval, and automatic summarization [1].

1.1 Challenges

In order to achieve optimal results, several significant challenges need to be addressed.

- Most of the data in this study is in English, while other languages are significantly under-represented. This poses challenges when it comes to analyzing and training the data effectively.
- In comparison to deep learning models, machine learning models perform worse on larger datasets.
- Getting a right number of same layers in CNN-LSTM hybrid model is a challenge, to maximize accuracy.

1.2 Contributions of the Paper

In this research, we suggest doing sentiment analysis with machine learning, deep learning, and pre-trained models and proposing a hybrid deep learning model.

- The paper utilizes various machine learning models, including The models utilized in this study include Random Forest, LGBM (LightGBM), XGBoost, Multinomial Naïve Bayes, AdaBoost, and Support Vector Machines (SVM). Their performance is assessed and compared using metrics such as accuracy and receiver operating characteristic (ROC) scores.
- The study employs advanced deep learning methodologies to construct a bidirectional model referred to as Long Short-Term Memory (LSTM). This LSTM model is built upon the principles of Recurrent Neural Networks (RNN). Particularly, the research utilizes a specific variant of RNN known as Bidirectional Long Short-Term Memory networks (BI-LSTM), which has showcased remarkable performance and surpasses various established models in effectively handling long-term dependencies.
- Additionally, it incorporates BERT, a pre-trained model that has shown superior performance compared to the aforementioned models utilized in this study.
- CNN-LSTM Hybrid model is trained to get a promising accuracy.

2 Literature Review

There are various innovative algorithms for language processing according to the dataset and problem statement. Plenty of research papers are there which has utilized different algorithms and created their own algorithm by stacking number of classifiers and tuning it accordingly. Some of them are discussed below:

2.1 Machine Learning Approach

Machine learning approaches can be broadly categorized into three types: supervised learning, unsupervised learning, and semi-supervised learning. In supervised learning, the documents are appropriately labeled, enabling the model to learn from labeled examples. On the other hand, unsupervised learning for text classification aims to identify and categorize texts that lack explicit labels or classifications [2].

Related Work in Sentiment Analysis Using Machine Learning Approach. Marta Fernandes, in paper [4], presented a novel method for assisting triage healthcare professionals in stratifying patients and recognizing patients at higher risk of ICU admission, data was broken down into train (70%) and test (30%) groups using Stratified random sampling and model training was done using 10-fold cross-validation. Joshua Acosta, in paper [5], Google's Word2Vec model was used to do sentiment analysis on twitter messages linked to US airline businesses. This model's word embeddings are often brought into practice to learn context and generate high-dimensional vectors in space, which are then categorised using machine learning techniques. Among the models used, Word2Vec achieved the highest accuracy of 72%. Following Word2Vec, the study employed support vector machines and logistic regression as the machine learning models, which yielded favorable results.

Arman S. Zharmagambetov [7] researched the use of Google's algorithm Word2Vec is a vector-based representation of words as fundamental linear algebra procedures that preserve semantic links between words. He concluded through his experiment that deep learning performed much better than Bag of Words model yielded marginally superior results. Monisha Kanakaraj, in paper [8], focused on increasing classification accuracy by the use of natural language processing (NLP) techniques such as semantics and word-sense disambiguation. Experiments have shown that the ensemble classifier outperforms traditional machine learning classifiers by a factor of three to five percent due to the high unpredictability in calculating divisions during feature vector subset selection. Mr. Jeevan Anil Phand, in paper [9], used twitter data to perform sentiment analysis and to do this, the tweets are first retrieved and classified using Stanford NLP as to whether they are positive, negative or neutral. Stanford NLP method performed better while predicting on certain datasets like India vs Pakistan (Match) which gives around 100% accuracy while for Amazon data, accuracy was 89% (Table 1).

As concluded from the above table, machine learning models are not able to capture better accuracy and some have no room for improving or tuning the model, we move towards deep learning which is capable and more successful in terms of training on large set of datasets.

Table 1. Summary of ML based papers

Authors	Methodology used	Classification technique	Result	Dataset used	Limitation in the paper
Fernande s, Marta & Sun, Haoqi & Jain, Aayushee & Alabsi, Sudeshna & Robbins, Gregory & Mukerji, Shibani & Westover, M Brandon 2021	LASSO regularization is applied for reducing some features. After applying LASSO, multiclass logistic regression model is trained on 70% data of dataset and then model performance is calculated using hold-out test dataset	LASSO model	The model attained a micro-average range of 0.98 beneath the receiver operating line and precision score of 0.81 in the testing set	Hospital Discharge summaries research	Result shows precision score of 0.81 which shows that data is not sufficient for model to perform better
Donald A. Szlosek, Jonathan M. Ferretti 2016	Support Vector Classifier, Decision tree classifier and KNeighbors Classifier is used for predicting the development of a minor traumatic brain damage in the absence of a manual review	Support Vector Machine, Decision trees and K Nearest Neighbour models	Out of all three models trained on the data, SVC performed best with 93.3% sensitivity and 97.62 specificity	3621 artificial patient's records receiving a computed tomograph y scan of the head for minor traumatic brain damage	External validation is required to replicate these findings and assess their performance in other contexts

(continued)

Table 1. (*continued*)

Authors	Methodology used	Classification technique	Result	Dataset used	Limitation in the paper
Joshua Acosta, Norissa Lamaute, Mingxiao Luo, Ezra Finkelstein, Andreea Cotoranu, 2017	Word2Vec is used to produce word embeddings which are further classified using machine learning algorithms	Word2Vec model	72% accuracy was accomplished through the use of a support vector machine and logistic regression model	US Airlines Sentiment	Different classifiers can be applied on word embeddings along with checking for content similarity
Arman S. Zharmagambetov 2016	This paper discusses the application of Word2Vec algorithm for text classification according to their sentiment types	Word2Vec model	Bag of words model gave 84.5% accuracy while deep learning gave 89.5% accuracy	50,000 IMDb movie reviews dataset from Kaggle	Error level in word's clustering may appear
Monisha Kanakaraj and Ram Mohana Reddy Guddeti, 2015	Ensemble learning models are used in this paper and compared to other machine learning algorithms	Ensemble Learning	Extremely randomized trees performed much better with accuracy of 92% because of high randomness in data selection	Tweets data fetched using twitter API v1.1	Some methods are used for increasing accuracy which can also be achieved with traditional algorithms by tuning them
Ms. Sheetal Anil Phand and Mr. Jeevan Anil Phand, 2017	Methodology is based on maximum entropy machine learning model	Machine learning model from Stanford NLP	Performance on various topics were evaluated and it performed well with 92% average accuracy among all	Tweets fetched using Twitter API	No room for improving or tuning the model

2.2   Deep Learning Approach

Deep Learning is influenced by the idea of our brain, which has multiple layers of perceptron i.e., artificial neuron. Long Short-Term Memory (LSTM) is a technique for learning from a word sequence. The LSTM is a Recurrent Neural Network (RNN) with more internal memory [3]. Various LSTM models outperform other approaches to detect negation in sentences. The brief architecture of LSTM is shown as follows (Fig. 1):

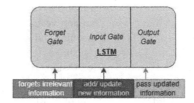

Fig. 1.   Depicts a simplified architecture of Long Short-Term Memory (LSTM).

Research in Sentiment Analysis Using a Deep Learning Technique Approach.   In a study conducted by Saeed Mian Qaisar [6], an LSTM classifier was employed to analyze the emotions expressed in IMDb movie reviews. The results revealed a remarkable classification accuracy of 89.9%, showcasing the potential integration of this suggested

Table 2.   Summary of DL based papers:

Authors	Methodology used	Classification technique	Result	Dataset used	Limitation in the paper
Saeed Mian Qaisar, 2020	Long Short-Term Memory (LSTM), model is applied	LSTM model	Best accuracy of 89.9% is achieved	Andrew Maas compiled 50,000 IMDb movie reviews	Additional data cleaning and the use of ensemble classification algorithms can improve accuracy
N. Sriram, 2019	This paper is focused on recurrent neural network language-based LSTM model which has certain advantages over RNN	LSTM model	Performance is of 92.8% accuracy on test or validation set of data and 95% accuracy on training set	US Airlines Sentiment	Third class i.e., neutral class can be added to the dataset in order to improve sentiment analysis

technique into existing text-based sentiment analysis methodologies. Similarly, N Sriram [16] proposed a language-based recurrent neural network model,

specifically utilizing LSTM, which effectively utilizes logic gates in its architecture to enable information retention and forgetting. The analysis was conducted on a dataset of US Airline reviews containing two classes, positive and negative (Table 2).

Despite the notable advancements of deep learning models over traditional machine learning approaches, there remains a potential for overfitting and subpar performance on test or validation datasets. In order to address or mitigate these challenges, researchers and practitioners have turned towards the utilization of hybrid neural network models.

2.3 Hybrid Model Approach

In CNN, max pooling layers are used for extraction of information and then reduced the feature dimensions by max pooling. Output of this is sent to LSTM network that decides which information should be remembered and which not to get the optimal results. In Fig. 2, CNN-LSTM architecture is shown for better understanding of the CNN-LSTM hybrid model (Table 3).

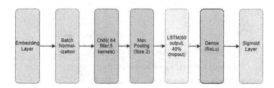

Fig. 2. Shows the brief architecture of LSTM-CNN Hybrid Model.

3 Methodology

3.1 Dataset and Pre-Processing

In our research, we made use of two datasets sourced from Kaggle: the Amazon Reviews Dataset and the IMDb 50K Review Dataset. The Amazon Reviews Data consists of a substantial collection of 34,686,770 reviews, featuring three distinct columns for text, sentiment, and title information. In our research, we focused on the text column, which contains the complete product review. The IMDb dataset consists of 49,582 user reviews on different movies, organized into two columns: review and sentiment. Both datasets are tasked with binary classification, as they involve distinguishing between two classes: positive and negative, represented by the labels 1 and 0 respectively. Despite the fact that reviews typically span a range of 1 to 5 stars for products or movies, the accompanying metadata for both datasets specifies that the reviews have already been categorized into the two aforementioned classes of positive and negative (Fig. 3).

Data Cleaning. The first stage in model training is data cleaning. The goal of this research is to cut out any extraneous words and expressions from texts to improve the machine learning model's performance.

Table 3. Related work in sentiment analysis using hybrid neural network model approach

Authors	Methodology used	Result	Dataset used	Limitation in the paper
Kathrein Abu Kwaik, Motaz Saad, Stergios Chatzikyriak idis, and Simon Dobnik, 2019	a deep learning model that combines CNN and LSTM layers was proposed, and it enhanced classification performance	The accuracy of binary classification improved from 81% to 93%, and accuracy of three-way classification improved up to 76%	LABR, ASTD and Shami-Senti(corp us)	Better accuracy can be achieved by further changing hyperparameters or tuning the layers
Abdallah Ghourabi, Mahmood A. Mahmood and Qusay M. Alzubi. 2020	A proposed detection model uses a combination of CNN- LSTM layers	CNN-LSTM model performs best among other models with 98.3% accuracy	Arabic SMS Dataset	Tuning hyperparameters can help increasing accuracy of the model

review	sentiment
I just adopted a chocolate lab who loves to sn...	1
Watched it and wasn't very impressed. It was t...	0
By 1967, enough was enough with the light fluf...	0
I was interested in what all the hype was abou...	0
I expected a well written book (as someone rec...	0

Fig. 3. Review Samples from Amazon Dataset.

The following items must be eliminated at this stage:

- Removed redundant punctuation.
- Conversion of capital words to lower words.
- Included the removal of HTML tags and specific emojis from the text data. Once the procedures are completed, the cleaned text data undergoes pre-processing.

Text Pre-processing. Text pre-processing is an essential phase in natural language processing, as machine learning models require numerical data rather than textual data for recognition and analysis. Therefore, it is imperative to transform the textual data into a suitable numerical format during this step. Lemmatization is performed. TF-IDF (Term Frequency-Inverse Document Frequency) is also performed after Lemmatization.

3.2 Ensemble Model

Ensemble methods or model stacking techniques are meta-algorithms employed to integrate diverse machine learning approaches, resulting in a consolidated predictive model.

The primary objective is to reduce variance, bias, or enhance prediction accuracy by leveraging the strengths of multiple specific models. This approach demonstrates superior performance compared to using a single model. In this study, the three highest-performing machine learning models, determined by their ROC scores after training on both datasets, are stacked together. The text column undergoes pre-processing and TF-IDF transformation before being used to train the stacked model. Finally, the performance of this ensemble model is evaluated on the test dataset (Fig. 4).

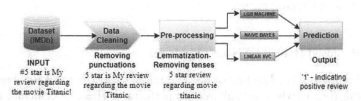

Fig. 4. Ensemble model of three most efficient machine learning models of IMDb Review dataset.

3.3 Implementation Using Bidirectional Long Short-Term Memory

In Fig. 5, Bidirectional LSTM is essentially an extension of Gated Recurrent Neural Network model. Bidirectional LSTMs are derived from bidirectional RNN [10], which analyses sequence data using two unique hidden layers, both forward and backward., as illustrated in the picture. The two hidden layers are linked to the same output layer by bidirectional LSTMs.

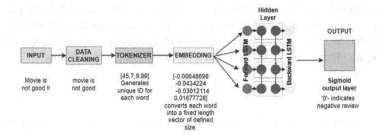

Fig. 5. Shows the architecture of Bidirectional Long Short-Term Memory (Bi-LSTM).

3.4 Implementation Using Bidirectional Encoder Representations from Transformers (BERT)

BERT, a widely-used pre-trained language model, offers contextual understanding of words based on its prior learning from an unannotated training dataset. Various versions of BERT exist, and this research paper focuses on utilizing the Distil-BERT base model. Distil-BERT is a condensed iteration of BERT that is designed to be smaller, lightweight, and faster, featuring a reduced parameter count compared to BERT base-uncased. It was pretrained keeping in mind the following:

Distillation Loss. The model was trained to produce the same results as the BERT basic model.

Masked Language Modelling (MLM). While considering a sentence, 15% of the input's words are covered and chosen at random by this. And then runs the entire masked text through the model. It gives the model an ability to become familiar with the sentence's two-way representation.

3.5 The Proposed Hybrid CNN-LSTM Model

IMDb review dataset is taken first of all. The initial stage of model training involves data cleaning, (Fig. 6) where several tasks are performed. These tasks include removing unnecessary punctuation, converting uppercase letters to lowercase, eliminating HTML tags and specific emojis from the text, and removing articles, conjunctions, and certain other elements from the text. Following data cleaning, the next step involves tokenization, which entails dividing the raw text into smaller units or segments. Once the text has been tokenized, the embedding layer is employed to transform each word into a vector representation of a predetermined size and length. To enhance stability and promote normalization within the layers, a batch normalization layer is employed. This layer rescales and recenters the inputs. Following this, a convolutional layer is utilized to extract relevant features. The resulting output from the CNN layers is then passed through a Max-Pooling layer, which reduces the dimensionality of the features. To obtain a sequence output instead of a single value, an LSTM layer is incorporated. In the final step, a dense layer Incorporating the Rectified Linear Unit (ReLU) activation function is included to enhance the generalization of the output from the LSTM layer. For the purpose of text message classification using our binary classification model, a dense layer with the Sigmoid activation function is utilized. Sigmoid function is a logistic function that ranges from 0 to 1, as defined by the formula given below-

$$y(x) = 1/(1 + e^{-x}) \tag{1}$$

CNN. The main purpose of the Convolutional Layer is to extract meaningful features from textual data. This is achieved by applying convolutional techniques on the word vectors produced by the embedding layer. The Rectified Linear Unit (ReLU) function is utilized as a nonlinear function in this process. It is written as below-

$$C(x) = max(0, x) \tag{2}$$

The ReLU activation function returns x if the value is positive and 0 otherwise.

Max Pooling. The Convolution technique is employed to generate feature maps that offer strong vector representation. Following the CNN layer, a Max-Pooling layer is incorporated to assist in selecting significant information by reducing the influence of weak activations. This step helps prevent overfitting caused by outlier text.

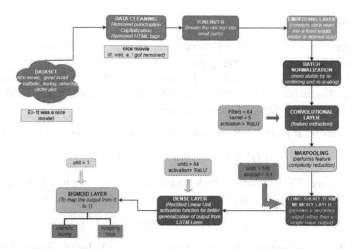

Fig. 6. Architecture of Proposed LSTM-CNN Hybrid model.

TM. A type of RNN that is capable of correlating recent and previous data is the LSTM. A memory cell mt and three gates—an input gate (ip), an output gate (op), and a forget gate (fg)—that regulate the flow of information within the LSTM unit make up each unit at time step t. These gates regulate the manner in which the current memory cell (mt) and the current hidden state (hd) are simultaneously changed. The following are the transition functions between LSTM units.: -

$$ip = \sigma(Wi \cdot [hd - 1 + bi]) \qquad (3)$$

$$fg = \sigma(Wf \cdot [hd - 1, xt] + bf) \qquad (4)$$

$$qt = tanh(Wq \cdot [hd - 1, xt] + bq) \qquad (5)$$

$$op = \sigma(Wo \cdot [hd - 1, xt] + bo) \qquad (6)$$

$$mt = fg \odot mt - 1 + ip \odot qt \qquad (7)$$

$$hd = op \odot tanh(mt) \qquad (8)$$

The input vector for the LSTM unit is 'xt', the sigmoid function is 'tanh,' the operator is 'element-wise product,' and the weight matrices and bias vector parameters, respectively, W and b, need to be learned during training.

3.6 Evaluation Metrics

Following training, the model's performance is evaluated using various performance metrics. Analysing machine learning models or algorithms is a requirement for any project. For testing a model, a variety of assessment tools are available. In this paper, models are evaluated based on accuracy and ROC score as accuracy metrics alone cannot identify the best model. Some of metrics are explained:

- Accuracy Score: It is the proportion of correct predictions to total predictions.

$$(TP + TN)/(TP + TN + FP + FN) \tag{9}$$

- Precision Score: The degree to which positive class predictions are actually positive class predictions is known as precision.

$$TP/(TP + FP) \tag{10}$$

where, TP = True Positive, TN = True Negative, FP = False Positive and FN = False Negative.

4   Results and Discussion

4.1   Results Obtained

Within the domain of machine learning, numerous ensemble models involving techniques for bagging and boosting are trained to do sentiment analysis. However, a modified version of the SVM classifier known as Linear SVC is employed in this study. Furthermore, the deep learning models yield encouraging outcomes, especially the BiGRU (Bidirectional Gated Recurrent Unit) model, a variant of the LSTM model, demonstrating remarkable accuracy. BERT model performed on Amazon Reviews dataset has shown better results than Bidirectional LSTM model with higher accuracy and ROC score. For IMDb Reviews dataset, Distil-BERT performed well with accuracy and ROC score which is better than previous Bi-LSTM model. During the "RMSProp" optimizer is chosen over the "RMSProp" optimizer for training the proposed model. Adam optimizer due to superior results achieved after conducting hyperparameter tuning. The Hybrid Model, on the other hand, is trained using the Amazon Reviews dataset. Which yielded an accuracy and ROC score which is best for this dataset of all the models trained. For IMDb Reviews dataset as well, hybrid model performed well with accuracy and ROC score best among all the models trained for IMDb dataset. The proposed model in this study offers a notable advantage in terms of speed compared to other models trained. This is primarily attributed to the inclusion of CNN-Max Pooling layers, which effectively perform feature extraction and reduction. This reduction in complexity of the input benefits the subsequent LSTM layers, where the utilization of logic gates allows for the retention and forgetting of information. The final accuracy and ROC scores for all the models are depicted in Table 4.

4.2   Findings Based on Models Used Machine Learning Results

Amazon Reviews Dataset. See Table 5.
   IMDb Reviews Dataset. See Table 6.

Table 4. Shows the Accuracy and ROC Curve Score of models for Amazon Reviews and IMDb Reviews.

S. No.	Models	Accuracy for Amazon Reviews dataset	ROC score for Amazon Reviews dataset	Accuracy for IMDb Reviews dataset	ROC score for IMDb Reviews dataset
1	Machine learning	85.2%	0.85	89.8%	0.89
2	Deep learning	88%	0.91	84%	0.91
3	BERT	85%	0.93	88%	0.95
4	Proposed model	88%	0.94	90%	0.96

Table 5. Shows the Accuracy, Precision Score and ROC Curve Score of Machine learning models for Amazon Reviews

S. No.	Models	Accuracy	Precision Score	ROC AUC Score
1	Decision Tree	0.710	0.713	0.710
2	AdaBoost	0.758	0.687	0.758
3	Gradient Boosting	0.762	0.793	0.761
4	Naive Bayes	0.819	0.830	0.819
5	XGBoost	0.824	0.823	0.824
6	LGB Machine	0.831	0.824	0.831
7	Random Forest	0.836	0.843	0.836
8	Linear SVC	0.851	0.858	0.851
9	**Ensemble**	**0.852**	**0.853**	**0.852**

Deep Learning Results. Accuracy and ROC curve for Amazon Reviews Dataset and IMDb Reviews Dataset are shown below (Fig. 7).

BERT Results. Accuracy and ROC curve for Amazon Reviews Dataset and IMDb Reviews Dataset are shown below (Fig. 8).

Proposed Model Results. Accuracy and ROC curve for Amazon Reviews Dataset and IMDb Reviews Dataset are shown below (Fig. 9).

Table 6. Shows the Accuracy, Precision Score and ROC Curve Score of Machine learning models for IMDb Reviews.

S. No.	Models	Accuracy	Precision Score	ROC AUC Score
1	Decision Tree	0.722	0.714	0.722
2	AdaBoost	0.806	0.834	0.805
3	Gradient Boosting	0.812	0.789	0.811
4	XGBoost	0.854	0.843	0.853
5	Random Forest	0.855	0.866	0.855
6	LGB Machine	0.862	0.857	0.862
7	Naive Bayes	0.864	0.881	0.865
8	Linear SVC	0.896	0.905	0.896
9	**Ensemble**	**0.898**	**0.902**	**0.898**

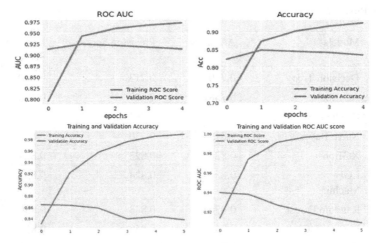

Fig. 7. Shows the ROC and Accuracy curve of Bi-LSTM Model for Amazon Reviews and IMDb Reviews respectively.

5 Conclusion and Future Works

Light Gradient Boosting Machine (LGBM), Naive Bayes, Random Forest, Linear Support Vector Classifier (SVC), Decision Tree, AdaBoost, Gradient Boosting, and Extreme Gradient Boosting (XGBoost) are some of the machine learning models that are taught. in this study. Two datasets, the IMDb review dataset and the Amazon review dataset, were

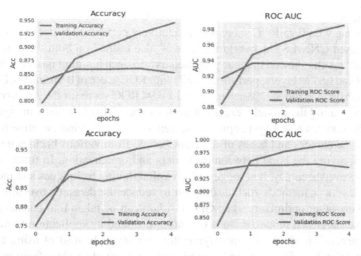

Fig. 8. Represents the Accuracy and ROC curve of BERT for Amazon Review and IMDb Reviews respectively.

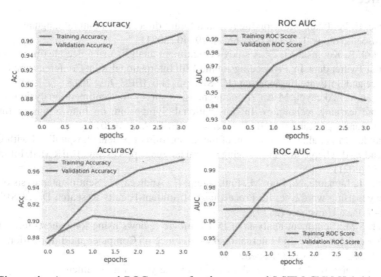

Fig. 9. Shows the Accuracy and ROC curves for the proposed LSTM-CNN Hybrid model for Amazon Reviews and IMDb Reviews respectively.

taken for diversity.. For attaining maximum accuracy among machine learning models, ensemble models for both datasets were trained which gave the accuracy of 85.2% and 89.8% respectively and ROC score of 0.852 and 0.898 respectively. Bidirectional-When trained on the Amazon review dataset, the LSTM model has an accuracy of 84% and a ROC score of 0.91. For IMDb review dataset also it gave 84.1% accuracy and 0.91 ROC score. Amazon review data trained on transformers-based Distil-BERT model produced the 85% accuracy and a ROC score of 0.93 for the IMDb dataset, it produced 88% accuracy along with 0.95 ROC score which relatively performed better than Bi-LSTM.

A proposed CNN-LSTM hybrid model was also trained on both the datasets. The experimental results showed that our proposed hybrid deep learning model, which combines CNN and LSTM layers, performed well, with ROC score of 0.96 and 90% accuracy for the IMDb dataset and 88.2% accuracy and a 0.94 ROC score for the Amazon reviews data, demonstrating the model's excellent performance. This solution can significantly improve the reviews given by people, by segregating positive and negative reviews to understand the choices and tastes of different people from various backgrounds which helps in improving the bond between customers and organization. In future works, to achieve improved accuracy, it is suggested to explore tuning techniques and incorporate additional layers or methods such as dropout to reduce the dataset's overfitting. Additionally, using other optimizers like SGD or Adam can be taken into consideration. In addition, the suggested model, there is potential to extend its application to multi-class sentiment analysis problems by modifying the final layer. Instead of using a sigmoid function, the SoftMax function can be employed for multi-class classification tasks.

References

1. Nehme, F., Feldman, K.: Evolving role and future directions of natural language processing in gastroenterology. Dig. Dis. Sci. **66**(1), 29–40 (2021)
2. Vateekul, P., Koomsubha, T.: A study of sentiment analysis using deep learning techniques on Thai Twitter data. In: Proceedings of the 13th International Joint Conference on Computer Science and Software Engineering (JCSSE), pp. 1–6. IEEE, July 2016
3. Aydoğan, E., Akcayol, M.A.: A comprehensive survey for sentiment analysis tasks using machine learning techniques. In: International Symposium on Innovations in Intelligent Systems and Applications (INISTA), pp. 1–7. IEEE, August 2016
4. Fernandes, M., et al.: Classification of the disposition of patients hospitalized with COVID-19: reading discharge summaries using natural language processing. JMIR Med. Inform. **9**(2), e25457 (2021)
5. Acosta, J., Lamaute, N., Luo, M., Finkelstein, E., Andreea, C.: Sentimentanalysis of Twitter messages using word2vec. In: Proceedings of Student-Faculty Research Day, CSIS, vol. 7, pp. 1–7. Pace University (2017)
6. Qaisar, S.M.: Sentiment analysis of IMDb movie reviews using long short- term memory. In: Proceedings of the 2nd International Conference on Computer and Information Sciences (ICCIS), pp. 1–4. IEEE, October 2020
7. Zharmagambetov, A.S., Pak, A.A.: Sentiment analysis of a document using deep learning approach and decision trees. In: Twelve International Conference on Electronics Computer and Computation (ICECCO), pp. 1–4. IEEE, September 2015
8. Kanakaraj, M., Guddeti, R.M.R.: Performance analysis of ensemble methods on Twitter sentiment analysis using NLP techniques. In: Proceedings of the 2015 IEEE 9th International Conference on Semantic Computing (IEEE ICSC 2015), pp. 169–170. IEEE, February 2015
9. Phand, S.A., Phand, J.A.: Twitter sentiment classification using Stanford NLP. In: Proceedings of the 1st International Conference on Intelligent Systems and Information Management (ICISIM), pp. 1–5. IEEE, October 2017
10. Nowak, J., Taspinar, A., Scherer, R.: LSTM recurrent neural networks for short text and sentiment classification. In: Rutkowski, L., Korytkowski, M., Scherer, R., Tadeusiewicz, R., Zadeh, L.A., Zurada, J.M. (eds.) ICAISC 2017. LNCS (LNAI), vol. 10246, pp. 553–562. Springer, Cham (2017). https://doi.org/10.1007/978-3-319-59060-8_50

11. Fernandes, M., et al.: Predicting intensive care unit admission among patients presenting to the emergency department using machine learning and natural language processing. PloS One **15**(3), e0229331 (2020)
12. Szlosek, D.A., Ferrett, J.: Using machine learning and natural language processing algorithms to automate the evaluation of clinical decision support in electronic medical record systems. eGEMs **4**(3), 5 (2016)
13. Abu Kwaik, K., Saad, M., Chatzikyriakidis, S., Dobnik, S.: LSTM-CNN deep learning model for sentiment analysis of dialectal Arabic. In: Smaïli, K. (ed.) ICALP 2019. CCIS, vol. 1108, pp. 108–121. Springer, Cham (2019). https://doi.org/10.1007/978-3-030-32959-4_8
14. Ghourabi, A., Mahmood, M.A., Alzubi, Q.M.: A hybrid CNN-LSTM model for SMS spam detection in Arabic and english messages. Future Internet **12**(9), 156 (2020)
15. Schuster, M., Paliwal, K.K.: Networks bidirectional reccurent neural. IEEE Trans. Signal Process **45**, 2673–2681 (1997)
16. Prabhakar, E., Santhosh, M., Krishnan, A.H., Kumar, T., Sudhakar, R.: Sentiment analysis of US airline twitter data using new adaboost approach. Int. J. Eng. Res. Technol. (IJERT) **7**(1), 1–6 (2019)

In-store Product Placement Using LiDAR-Assisted Discrete PSO Algorithm

Shajulin Benedict[(✉)] [iD], R. Lakshin Kumar [iD], and M. Baranidaran [iD]

Indian Institute of Information Technology Kottayam, Valavoor P.O.,
Kottayam District 686635, Kerala, India
{shajulin,lakshin2019,baranidaranm20bcs40}@iiitkottayam.ac.in
https://www.sbenedictglobal.com

Abstract. The fundamental reason for a business failure is not only because of the quality of products. Novel IoT-enabled approaches have good reasons to proliferate businesses in multitudes using IoT-enabled solutions. The traditional approach of placing products in pre-identified positions in shopping malls or showroom centers has delayed or missed customer attractions. This article proposes a Light Detection and Ranging (LiDAR)-based in-store product placement approach. The proposed mechanism utilized a LiDAR sensor to quickly identify the missing products from racks; it incorporates Discrete Particle Swarm Optimization (Discrete PSO) algorithm-specific product placement mechanism. The decisions of Discrete PSO attempted to maximize the attraction of customers and minimize the cost involved in transporting products to corresponding racks. The proposed approach was experimented with at the IoT Cloud Research laboratory. The experimental results reveal that the Discrete PSO algorithm-based product placement approach reduced the delivery time of products and increased customer attraction with an efficiency of 38.02 percent in combined objectives. The approach will be beneficial for increasing the number of customers and product sales in in-store business locations.

Keywords: Edge Computing · Discrete PSO · In-store · IoT · Product Placement

1 Introduction

In-store product placement puts the focus on the struggle within the supermarket or product organizers of malls for attaining improved businesses over branding intelligence. Specifically, the buying habits of customers increase when products are attractively aligned even during busy business hours [6]. Also, the visual impact of products that are placed in stores attracts customers owing to the consistent views of product placements.

Supported by IIIT-Kottayam and its labs, including AIC-IIITKottayam.

Traditionally, in-store product placement was based on manual methods [13]. The shopping mall's product organizer has to align products on pre-defined locations of branded products' racks before any potential customers visit the vicinity. The inclusion of IoT-enabled solutions based on sensors and actuators has increased the business by attracting customers on a large scale in recent years. However, there are several notable challenges in the existing in-store product placement solutions. They are listed as follows:

1. *Huge Shopping Malls* – in the majority of smart cities, in recent years, the size of shopping malls or retail shops has tremendously increased;
2. *Increased Number of Customers* – people continue pursuing offline shopping, especially on holiday seasons such as Christmas, Diwali, and so forth, in large numbers;
3. *Speedy Awareness and Speedy Placements* – IoT-enabled solutions such as capturing the current availability status of products on racks or informing concerned product organizers have potential business improvement options. However, placing the products on empty slots of racks and positioning product floors [19] before the visit of customers or based on the prediction of customer's arrival is a challenging aspect due to the existing poor automation strategies; and,
4. *Inefficient Product Transportation* – deciding on the transportation of products based on the need of customers or the requirement of products on racks of shopping malls [2,13] using sophisticated scheduling algorithms is not widely discussed in the past.

Obviously, there is a dire need for an efficient product placement approach that collects the updated information of empty product locations on racks and suggests a cost-efficient transportation of products – i.e., the IoT-enabled solutions must be intelligent enough to handle the business portfolio considering several factors, including customer satisfaction, at ease.

In this article, we proposed a LiDAR-assisted in-store product placement approach using the Discrete PSO algorithm. The LiDAR sensor arrays are utilized in the architecture to collect the empty slots of products periodically. Based on the input sensed data, the Discrete Particle Swarm Optimization algorithm is applied to find the best path to deliver products considering the frequency of customers (who collect products) and the time required to deliver products on the missing positions of racks. The proposed approach is suitable to increase the potential customers and business opportunities in shopping malls.

To manifest the significance of the proposed approach, we organized experiments at our IoT Cloud Research laboratory. Our experiments revealed an efficiency of over 38 percent with combined objectives of maximizing customers and minimizing transportation time of products to racks when applied using the proposed Discrete PSO algorithm and LiDAR sensor-based input sources.

The rest of the article is described as follows: Sect. 2 expresses the state-of-the-art work on the in-store product placement domain; Sect. 3 explains the proposed architecture to collect sensed input data and apply intelligence using

Discrete PSO algorithm; Sect. 4 details on the steps and processes involved in the Discrete PSO algorithm; Sect. 5 manifests the importance of the proposed method using sufficient experimental results and discussions; and, finally, Sect. 6 expresses a few conclusions and near-future outlooks.

2 Related Works

The growth of IoT-enabled sensor-based distributed applications as an alternative to the traditional concept of centralized computational services is built around its recent advances in micro-controller-based technologies and computational capabilities of tiny machines, including edge-enabled sensor devices. Tens of thousands of applications have evolved in recent years in various domains such as agriculture, healthcare [7,14], industries, retail sectors [11], finance, education [9], and so forth.

Among many available sensors such as temperature sensors, humidity sensors, and healthcare-related sensors, the LiDAR sensor is most widely applied in tracking objects or positioning objects in 360-degree views. For instance, authors of [17] have devised a railway crossing system using 3D-based LiDAR sensors to more accurately detect objects; authors of [1] have proposed the application of a LiDAR-based IoT-enabled system to automatically detect the position of objects in a car movement lane; and, so forth.

In the store-based product placement domain, a few research works have been carried out in the past to study human behavior in consuming products. For instance, authors of [15], have studied customer behavior using beacons to develop personalized recommendations to Brick-and-Mortar stores; authors of [12] have developed a group recommendation system on offline product store using a machine learning-based collaborative filtering approach; authors of [13] have manually observed the product placement and rack positioning methods considering the branding and customer behaviors. Similarly, in the past, authors of [2] have developed a PrefixSpan algorithm to place products on racks considering the consumers' buying patterns. However, the authors have not applied sophisticated sensor-enabled techniques to quickly identify the empty racks of products.

A few authors of [4] have developed a generalized utility itemset (GUI) index that improves the revenue of retailers considering the sale pattern of in-store products. Recently, authors of [16] have developed a Particle Swarm Optimization (PSO) algorithm to guide an unmanned vehicle during the product delivery service. However, this work has not included the availability of customers to prioritize the importance of delivering products to appropriate lanes.

In general, heuristic methods such as PSO, Genetic Algorithm (GA), Ant Colony Optimization (ACO), and so forth, have been applied in several placement problem domains. Authors of [5] have utilized the PSO algorithm to optimally place gateways to improve services; authors of [8] have developed a combination of Discrete PSO and GA parameters to optimize the placement of data while distributing workflow tasks; authors of [10] have applied Whale Optimization Algorithm (WOA) to identify the placement of motion sensors considering several parameters of a smart home. A few authors have studied the application of meta-heuristic algorithms to plan the flow of manufacturing units considering various parameters such as finish time and material availability. Notably, authors of [3] have developed an energy-aware scheduling approach to plan the flow of completing the development of products in various manufacturing units.

Although a few works exist in different product placement domains, as it could be observed, the works have either not extensively studied the placement of products using optimization algorithms or have not discussed the placements considering the availability of customers. This work applied the concept of customer availability and placement of products in shopping malls using the Discrete PSO algorithm.

3   LiDAR-Assisted In-store Product Placement

In-store product placement on time is crucial for attracting customers and increasing business. This section explains our proposed method of applying LiDAR-assisted product placement. It covers the designed architecture and the functional capabilities of the architecture, including the entities of the architecture.

3.1   Architecture

The proposed architecture involves a few components such as Product Position Indicator (PPI), Discrete PSO sequencer, Customer Analyzer, Product/Rack Analyzer, Autonomic Actuator, and User Interface (UI). The major activities of these components are described below:

Product Position Indicator (PPI). Products are positioned on racks in in-store shopping malls for attracting customers considering several aspects. The costly products are often kept on the top portion of racks when compared to the frequently purchased products. The brands are often aligned without gaps so that customers visiting the product lane are attracted to purchase them.

Once when the products are removed from a rack by any consumer, the position of the product needs to be observed automatically so that the empty gaps could be replaced with similar products in the same position quickly before another customer will reach the vicinity.

To do so, in this article, we have included a LiDAR-based sensor-assisted Discrete PSO algorithm that suggests filling up the products in a huge shopping mall. The LiDAR sensors are fitted in arms that rover around the racks at different positions to collect the empty slots in a periodic fashion. LiDAR sensor is a sensor that measures the distance of objects at a 360-degree angle. The purpose of choosing a LiDAR-based indicator is to capture the entire map of the vicinity which includes the gaps of products on a rack and customer availability. The collected sensed information is transported to edge nodes for processing the position of products and for identifying product IDs.

Edge Nodes. The proposed architecture involves edge nodes such as raspberry pi, JetsonNano, or single-server units to pre-process the data so that the exact position of the products and their corresponding product IDs are collected. In addition, edge-level nodes are applied to extract the information of potential customers of a product from LiDAR's sensed data. For instance, the amount of time a customer stays nearer to a product indicates the chances of purchasing the product. However, it is crucial to relate customer behavior concerning the products purchased in a shopping mall. Hence, the position of products and customer availability-related information are pre-processed in edge nodes before they are transported to cloud services.

Gateways. Gateways are utilized in the architecture to transfer data between edge nodes and cloud services. They are important in the architecture to connect multiple communication protocol-enabled LiDAR sensors or edge devices. For instance, LiDAR sensors are manufactured to support 6LowPAN protocol, WiFI, or Bluetooth communication protocols. These communication protocols need to be supported by higher computational machines through gateways so that the connected cloud services are supported with ease.

Cloud Services. The product positions and customer availability information are forwarded to cloud services to plan logistics in transferring the empty products of racks from central storage units of shopping malls to the corresponding racks. The services involved in the proposed architecture include:

1. Discrete PSO Sequencer – The purpose of Discrete PSO Sequencer is to collect the rack IDs and product IDs that are empty and provide a logistic plan considering lane distance and customer availability. The racks might need to display certain products immediately owing to the availability of customers and non-availability of products. In some situations, due to the non-availability of customers, the products need not be delivered in appropriate racks. The Discrete PSO Sequencer applies the Discrete PSO algorithm while identifying the flow of filling up products in racks in shopping malls.
2. Analyzer – Depending on the number of purchases, the branding quality of the product is assessed using Analyzers. The services relating to the branding processes are quite often important in a business era.

User-Interface. The users of the solution involve shopping mall authorities or product distributors. The user-interface offers synchronous access to users while asynchronously connecting with communication nodes, including edge devices. In addition, the interface provides appropriate notifications when system fails.

4   Discrete PSO–Algorithm and Stages

The Discrete PSO algorithm is utilized in the proposed architecture to plan the flow of the delivery of products based on the gaps found in racks, customer

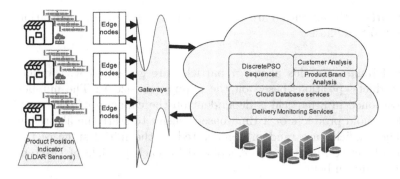

Fig. 1. In-Store Product Placement Architecture using Discrete PSO

availability, and the distance between the distribution unit and the location. This section explains the internal details of the proposed Discrete PSO algorithm to the product placement problem, the objective functions involved, and the stages of Discrete PSO algorithmic executions.

4.1 Algorithm

In general, the PSO algorithm is based on the swarm's intelligence or behavior while searching for its prey. It is a nature-inspired scheduling algorithm [18] based on a bird's flocking behavior. It is widely applied in several scheduling contexts of multiple domains, including traveling salesman problems. The PSO algorithm utilizes particles that represent the swarms flocking over the air. It is similar to other evolutionary techniques, including genetic algorithms. However, differences exist in terms of applying the algorithm to different optimization problems.

In our problem domain, an entire particle represents a sequence of product racks' positions that need to be placed while they are emptied by customers. For instance, if a product is purchased by a customer, the rack of the product becomes empty. This will be notified by LiDAR-assisted devices and the sequence of product racks that need to be filled up with newer products to attract customers is identified considering the number of customers and the distance between the racks and distribution units. They are typically represented as follows:

```
Sequence (Particle's Positions):
    PP1 PP6 PP10 PP2 PP8 PP3 PP7 PP5 PP4 PP3
```

Varying the order of serving products positions, the distance to reaching racks and products differ. Each particle, therefore, corresponds to a sequence of products' positions.

The major parts of the algorithm are classified into two phases – initialization phase and the iterative phase.

Initialization Phase – In this phase, n particles are generated with different order of products' positions based on the population size P. The number of products positioned in a particle is dependent on the capacity of the transporting vehicle to hold products or a time-based trigger to start the algorithm periodically. The particles are randomly generated at the initial stage of the algorithm. The objective functions are calculated for each particle of Discrete PSO algorithm as shown below:

$$Objective\ 1\ (Obj1): \tag{1}$$
$$Minimize \Rightarrow Total\ Distance\ (TD) = \Sigma_{i=0}^{m}(d_i)$$

where m resembles the number of products' positions where the products have to be positioned; d corresponds to the distance required by the transporting vehicle to reach the location where the products are empty.

$$Objective\ 2\ (Obj1): \tag{2}$$
$$Maximize \Rightarrow Customer\ Availability\ (CA) = \Sigma_{i=0}^{m}(C_i) * p$$

where C resembles the customers identified nearer to the products' position that has to be served to increase the business possibilities; p corresponds to preferences. The value of p is higher when i is lower than half the value of m. This provides more preferential weights to the Discrete PSO algorithm while serving products in positions where customers are queued up rather than planning the logistics where the customers are not available.

$$Combined\ Objectives\ (CO): \tag{3}$$
$$Minimize \Rightarrow CO = w1 * TD + 1/(w2 * CA)$$

where, w1 and w2 are equal weights corresponding to this problem. In this work, we have considered w1 and w2 as 0.5.

In addition to identifying the Combined Objectives of each particle, they are assigned with random initial velocity v_i. Also, a variable that collects the local best value LB_p and global best value GB_p for the entire initial population is recorded in a list along with their corresponding particles.

Update Phase – During the update phase, the position and velocity of particles are iteratively modified similar to birds flocking for any prey. In Discrete PSO, the position is updated by shifting the current position of serving racks in a product delivery sequence. For instance, if the velocity is (2,3), (5,6), the sequence of racks in a particle will be modified as given below:

```
Sequence of Product Positions (PP)
-- PP1 PP6 PP10 PP2 PP8 PP3 PP7 PP5 PP4 PP3
After Applying Velocity (2,3)
-- PP1 PP10 PP6 PP2 PP8 PP3 PP7 PP5 PP4 PP3
After Applying Velocity (5,6)
-- PP1 PP10 PP6 PP2 PP3 PP8 PP7 PP5 PP4 PP3
```

Accordingly, the updated position of each particle is given as follows:

$$PP_i(t+1) = PP_i(t) + v_i(t) \tag{4}$$

Also, the velocity of each particle is updated by an equation given below:

$$v_i(t+1) = c_1 * rand_1 * v_i(t) + c_2 * rand_2 * (^eRS_i^t - RS_i^t) \tag{5}$$
$$c_3 * rand_3 * (GB - RS_i^t)$$

where, c_1, c_2, and c_3 are constants; $rand_1$, $rand_2$, and $rand_3$ are random numbers that range between 0 and 1.

5 Experimental Results

This section explains the experiments carried out at the IoT Cloud Research Laboratory of our premise to demonstrate the proposed in-store product placement approach. Initially, the experimental setup is explained; next, the identification of products' positions in racks using a LiDAR-based sensor is illustrated; and, finally, the importance of applying a Discrete PSO algorithm to quickly identify the sequence of placing products on racks is explained.

5.1 Experimental Setup

Experiments were carried out such that LiDAR-based sensors collect the position of products on racks. Any gaps in the pre-identified positions of products on racks are identified and the corresponding product positions, including the product IDs, are collected. For instance, PP23 or PP456 corresponds to the product's position which is numbered 23 or 456. This resembles the in-store product placement scenario of a shopping mall. In addition, the customers staying nearer to the product's position are identified using the LiDAR sensor's 360-degree position monitoring method.

The number of products that could be transported by a vehicle to different rack positions is defined before the experiments. In this work, we have not utilized vehicles to automatically transfer the products to racks; instead, we attempted to identify the products that are emptied in a rack and the sequence of the total number of products that need to be transported to racks considering the distance and the availability of customers. Additionally, we have assumed that all identified numbers of empty products using LiDAR sensors are scheduled using the Discrete PSO algorithm and transported to the respective positions for a specified time.

5.2   LiDAR-Based Data Collection

To mimic the situation of monitoring LiDAR-based products' position identification, we set up a product exhibition rack and placed the LiDAR sensor connected to the edge-based server node (see Fig. 2). Although the products were evenly placed, some products were mandatorily removed from the intended positions that represent the product's position. Similarly, a few obstacles were placed in front of these products to resemble the position of customers. The LiDAR sensors revolved 360-degree to calculate the distance d of objects in centimeters for each θ value ranging from 0 to 360 °C. The collected data were filtered in the form of product positions (PP) and the number of customers (nC) before they were submitted to the Discrete PSO Sequencer.

Fig. 2. LiDAR-based Product's Position Observation

5.3   Discrete PSO Sequencer

The role of the Discrete PSO sequencer is to identify the sequence of serving empty product positions in a shopping mall considering the objective functions. In this work, the Discrete PSO algorithm is evaluated using the parameters listed in Table 1. The parameter values are chosen based on a trial-and-error approach that offered a better CO efficiency.

The number of products resembles the product identification in shopping malls; whereas, the number of product gaps resembles the missing products as customers would have purchased them. The number of product gaps is defined based on the position of products in racks. Although the parameters could be fine-tuned, we experimented with the algorithm based on these settings to evaluate the capabilities of the proposed algorithm.

Table 1. Discrete PSO Parameters

Discrete PSO parameters	Values
Number of Populations	50
Number of Timesteps	100
Maximum No. of Products	1000
Maximum No. of Product Gaps	20
Constant c1	1
Constant c2	2
Constant c3	2
rand 1	0.2
rand 2	0.3
rand 3	0.4

Global Best Solution – Product Delivery Sequence. We studied the flow of product delivery on different racks in mocking in-store shopping malls using Discrete PSO algorithm. The algorithm identified the local best and global best solutions – i.e., the product delivery sequence, as shown in Table 2. We conducted four different experiments using varied product delivery requirements. The sequence of delivery considering the customers' availability and emptiness in products' locations is studied for 5 to 20 identified product positions. The identified flow of delivering the products and the corresponding CO are expressed in Table 2.

The combined objectives were observed as 266, 402, 1142.5, and 1651 for product positions 5, 10, 15, and 20. These values were obtained when the population size of the Discrete PSO was set as 50 and the number of time steps of the algorithm as 100. It could be observed that the CO values increased drastically when we had to identify more product's positions. For instance, there is a difference of 135 units while observing the product positions for 5 and 10 – i.e., the difference between CO values when experimented with 10 and 5 product positions is 135 units. On the contrary, we could observe that the CO values increased to 740.5 and 508 units between 10, 15, and 20.

Evaluation – CO Efficiency and Discrete PSO Stages. In addition to identifying the flow of delivering products to fill up the empty slots in racks, we evaluated the capability of the Discrete PSO algorithm. To do so, we observed the improvements in Discrete PSO solutions over a period of time. Figure 3 expresses the value of CO obtained while identifying the flow of delivering products in racks.

As observed, the CO value reduced from 2224.5 to 1104.5 while increasing the time steps of the algorithm. However, after 1000 time steps, the CO value was constant for almost 10000-time steps. Notably, the update phase time of the algorithm abruptly increased – i.e., for 1000 time steps, the update phase time was 8.116 s.; for 100000-time steps, the update phase time was 790.27 s.

Table 2. Global Best Solutions – Product Delivery Sequence using Discrete PSO Algorithm

Product Position	CO	Global Best
5	266.033	PP383 PP777 PP793 PP886 PP915
10	402.002	PP421 PP383 PP335 PP386 PP492 PP649 PP886 PP915 PP793 PP777
15	1142.5	PP763 PP915 PP886 PP777 PP649 PP383 PP421 PP335 PP386 PP27 PP59 PP362 PP690 PP793 PP492
20	1651	PP172 PP763 PP736 PP886 PP915 PP926 PP27 PP59 PP386 PP421 PP362 PP690 PP777 PP793 PP649 PP426 PP383 PP335 PP540 PP492

We achieved an average CO efficiency of 38.02 percent when compared to 10-time steps in larger time steps. Also, we noticed that the time required for executing the initialization phase of the Discrete PSO algorithm was almost constant for all the time steps of the algorithm.

The limitations are listed as follows: i) this work applied the Discrete PSO algorithm for identifying the product placements in in-store malls. However, it could be extended with multiple other heuristic algorithms; ii) a comparative study on applying multiple sensors with different heterogeneous characteristics is not explored.

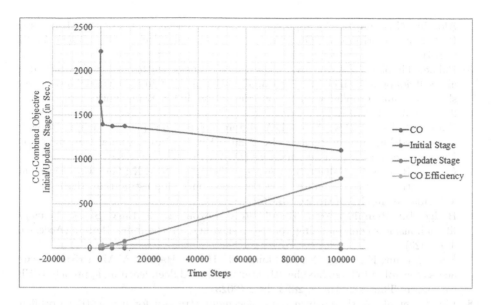

Fig. 3. Evaluation of CO Efficiency in Discrete PSO Stages

6 Conclusion

The application of IoT-enabled solutions in varied sectors, including retail sectors, has tremendously increased in recent years. The in-site product placement problem, which can be applied in automated shopping malls, is addressed in this article using LiDAR-assisted Discrete PSO algorithmic solution. The approach considered customer availability and product non-availability to schedule a plan for delivering products to appropriate locations using the Discrete PSO algorithm. Experiments were carried out at the IoTCloud Research Laboratory to study the importance of applying a Discrete PSO algorithm using varying time steps and the other Discrete PSO parameters. Our approach identified a flow of delivering products on racks of shopping malls with an efficiency of achieving over 38 percent combined objective value. In the future, the algorithm could be deployed in leading shopping malls; the proposed approach could have experimented with the other meta-heuristic algorithms that are supported by machine learning models.

Acknowledgements. The authors thank the support provided by IIIT-Kottayam through its labs, including the AIC-IIITKottayam, for completing the tasks.

References

1. Anand, B., Barsaiyan, V., Senapati, M., Rajalakshmi, P.: Region of interest and car detection using lidar data for advanced traffic management system. In: 2020 IEEE 6th World Forum on Internet of Things (WF-IoT). IEEE (Jun-2020)

2. Aloysius, G., Binu, D.: An approach to products placement in supermarkets using PrefixSpan algorithm. J. King Saud Univ. - Comput. Inform. Sci. **25**(1), 77–87 (2013)
3. Babaee Tirkolaee, E., Goli, A., Weber, G.W.: Fuzzy mathematical programming and self-adaptive artificial fish swarm algorithm for just-in-time energy-aware flow shop scheduling problem with outsourcing option. IEEE Trans. Fuzzy Syst. **28**(11), 2772–2783 (2020)
4. Bapna, C., Reddy, P.K., Mondal, A.: Improving product placement in retail with generalized high-utility itemsets. In: 2020 IEEE 7th International Conference on Data Science and Advanced Analytics (DSAA). IEEE (Oct 2020)
5. Guo, J., Rincon, D., Sallent, S., Yang, L., Chen, X., Chen, X.: Gateway placement optimization in LEO satellite networks based on traffic estimation. IEEE Trans. Vehicular Technol. **70**(4), 3860–3876 (2021)
6. Hodge B.: Planning Your Retail Store Layout in 8 Easy Steps. https://fitsmallbusiness.com/planning-your-store-layout/, in Planning Retail, (Accessed Jan 2023)
7. Lin, H., Kaur, K., Wang, X., Kaddoum, G., Hu, J., Hassan, M.M.: Privacy-aware access control in IoT-enabled healthcare: a federated deep learning approach. IEEE Internet of Things J. **10**(4), 2893–2902 (2023)
8. Lin, B., et al.: A time-driven data placement strategy for a scientific workflow combining edge computing and cloud computing. IEEE Trans. Indust. Inform. **15**(7), 4254–4265 (2019)
9. Liu, X.: Blockchain-enabled collaborative edge computing for intelligent education systems using social IoT. Inter. J. Distrib. Syst. Technol. **13**(7), 1–19 (2022)
10. Nasrollahzadeh, S., Maadani, M., Pourmina, M.A.: Optimal motion sensor placement in smart homes and intelligent environments using a hybrid WOA-PSO algorithm. J. Reliable Intell. Environ. **8**(4), 345–357 (2021)
11. Neunteufel, D., Grebien, S., Arthaber, H.: Indoor positioning of low-cost narrowband IoT nodes: evaluation of a TDoA approach in a retail environment. Sensors **22**(7), 2663 (2022)
12. Park, J., Nam, K.: Group recommender system for store product placement. Data Mining Knowl. Dis. **33**(1), 204–229 (2018)
13. Sigurdsson, V., Saevarsson, H., Foxall, G.: Brand placement and consumer choice: an in-store experiment. J. Appli. Behav. Anal. **42**(3), 741–745 (2009)
14. Benedict, S.: IoT-Enabled remote monitoring techniques for healthcare applications - An Overview. Informatica J. **46**, 131–149 (2022). https://doi.org/10.31449/inf.v46i2.3912
15. Voelz, A., Mladenow, A., Strauss, C.: Beacon technology for retailers - tracking consumer behavior inside brick-and-mortar-stores. In: Strauss, C., Kotsis, G., Tjoa, A.M., Khalil, I. (eds.) DEXA 2021. LNCS, vol. 12923, pp. 380–390. Springer, Cham (2021). https://doi.org/10.1007/978-3-030-86472-9_35
16. Watanabe, M., Ihara, K., Kato, S., Sakuma, T.: Introducing an AGV system into the warehouse and optimizing product placement for efficient operation. In: 2022 IEEE 11th Global Conference on Consumer Electronics (GCCE). IEEE (18 Oct 2022)
17. Wisultschew, C., Mujica, G., Lanza-Gutierrez, J.M., Portilla, J.: 3D-LIDAR based object detection and tracking on the edge of IoT for railway level crossing. IEEE Access **9**, 35718–35729 (2021)

18. Zhi, X.H., et al.: A discrete PSO method for generalized TSP problem. In: Proceedings of 2004 International Conference on Machine Learning and Cybernetics (IEEE Cat. No.04EX826). IEEE (2004)
19. Xie, Z.: TransFloor: Transparent floor localization for crowdsourcing instant delivery, Proc. ACM Interact. Mob. Wearable Ubiquitous Technol. **6**(4) 189, 1–30 (2023). https://doi.org/10.1145/3569470

Pattern Recognition

Deep Residual Variational Autoencoder for Image Super-Resolution

Justice Kwame Appati ⃝, Pius Gyamenah ⃝, Ebenezer Owusu$^{(\boxtimes)}$ ⃝,
and Winfred Yaokumah ⃝

Department of Computer Science, University of Ghana, Accra, Ghana
{jkappati,ebeowusu,wyaokumah}@ug.edu.gh,
pgyamenah001@st.ug.edu.gh

Abstract. Generating a higher version from a low-resolution image is a challenging computer vision task. In recent studies, the use of generative models like Generative Adversarial Networks and autoregressive models have shown to be an effective approach. Historically, the variational autoencoders have been criticized for their subpar generative performance. However, deep variational autoencoders, like the very deep variational autoencoder, have demonstrated its ability to outperform existing models for producing high-resolution images. Unfortunately, these models require a lot of computational power to train them. Based on variational autoencoders with a custom ResNet architecture as its encoder and pixel shuffle upsampling in the decoder, a new model is presented in this study. Evaluating the proposed model with PSNR and SSIM reveal a good performance with 33.86 and 0.88 respectively on the Div2k dataset.

Keywords: Super-Resolution · Variational Autoencoder · Upsampling · Image Resolution · Adversarial Networks

1 Introduction

In computer vision, SISR task deals with how to generate HR-images when given LR-images. The goal is for the resultant image to have either good or much better resolution when compared visually or objectively to the original image data. The SISR problem is described as difficult and ill-posed since a single LR image can yield several HR images (Dong et al. 2016). Application domain for image super-resolution is enormous. Among them are medical and satellite imaging, digital forensics, and surveillance among others where images of high quality are required. Various techniques have been explored in the literature to improve image resolution across various domains. Earlier approaches used classical computer vision approaches such the sparse coding techniques (Gu et al. 2015), edge-based approaches, statistical methods, and patch-based techniques (Wang et al. 2021). More recently, NN approaches have been used in the SISR task to mimic or replace the classical approaches. Advances in deep learning research and the development of GPUs and high-performance computers have laid the foundation for more experiments and research on DL-based SISR.

J. Abawajy et al. (Eds.): ICICCT 2023, CCIS 1841, pp. 91–103, 2023.
https://doi.org/10.1007/978-3-031-43838-7_7

In the study of He et al. (2015) and Krizhevsky et al. (2017), DL models such as CNNs, are widely used in SISR due to their effectiveness in image classification and other tasks like object and facial recognition. They also received more attention as result of their effectiveness and simplified implementation on modern GPUs. Among these variants of CNN as noted in Dong et al. (2014, 2016) are FSRCNN and SRCNN. In the case of generative adversarial network-based methods we have the ESRGAN as found in Sarode et al. (2022) used in SISR task. Using subjective qualitative assessment techniques such as SSIM, IFC and PSNR on the stated existing methods shows promising results. Again, CNN approaches have also shown improvement in hardware utilization regarding memory and CPU usage. While these deep learning-based SR approaches have proved useful and effective across different domains, more work is still being done to further increase their performance.

Recent work has shown the potential of generative models such as GANs (Goodfellow et al. 2014) and autoregressive models for image generation (Brock et al. 2018; Uria et al. 2016; Chen et al. 2018). These generative functions have also yielded comparable results in SISR task. GAN-based approaches like the SRGAN (Ledig et al. 2016) and the ProSRGAN (Wang et al. 2018) have yielded impressive results on benchmark datasets and evaluation criterion. VAEs have also shown progress in the image generation task. However, they have been criticized for having low generative capabilities when it comes to the super-resolution task (Chira et al. 2022). However, deeper variants of variational autoencoders such as the VDVAE (Child et al. 2020) have demonstrated that VAEs can be effective at the SISR task. A problem with deep VAEs such as the VDVAE is that they are difficult to train and use a lot of computational power. The effectiveness of variational autoencoders is influenced by the choice of the encoder and decoder design, loss function as well as how the latent representations are produced. The effective use of VAEs for the super-resolution remains an ongoing research task as Chira et al. (2022) alluded to.

In this paper, we further explore the effectiveness of VAEs in the SISR task by borrowing ideas from good-performing SISR models into the encoder and decoder network of a vanilla VAE. We experiment with different sizes of the latent dimension and assess our findings using the div2k benchmark dataset. Our work exhibits an impressive performance with reference to PSNR and SSIM.

2 Related Work

2.1 Conventional SR Methods

Until DL-based SISR techniques were developed, classical computer vision approaches were applied to the task. Earlier models used approximation algorithms such as cubic convolution interpolation (Keys 1981) to estimate or predict pixel values. Irani and Peleg (1991) reconstructed HR images from LR images by approximating the scenes from the ground truth image iteratively by simulating the imaging process, comparing the simulated image to a ground truth, and using back projection to minimize an error function.

Other approaches directly extract patches from input images and use them for upscaling. In the work of Freedman and Fattal (2011), higher resolution images are obtained without using example images or back-projection equations. They argue that the prediction-based methods such as the cubic and bicubic approaches are problematic since natural images may contain edges and other characteristics that introduce discontinuities and hence violate the underlying assumption of "analytical smoothness" that underlines these approaches. They assume self-similarity on natural images and use patches extracted from extremely localized parts of the input image rather than leveraging a database of examples. Their method yielded encouraging results for small scaling factors, where there are more examples of patches of greater significance.

2.2 DL-Based SR Models

Dong et al. (2016) played the pioneering role of using CNN for the SISR task. They suggested a DL-based SR model that made use of mapping LR images onto the HR images. This approach replaced the traditional method of utilizing a manual dictionary mapping in sparse-coding techniques with a deep learning-based method. The authors proved that their deep learning technique was equivalent to the traditional sparse-coding approach, providing justification for using hidden layers to obtain the mappings from LR images onto HR images rather than a dictionary. Using traditional computer vision techniques, they upsampled the input images to coarse HR images of an appropriate size and sent them into a feedforward CNN to reconstruct the HR output from the input. The resulting SRCNN model (Dong et al. 2016) was shown to produce HR images from LR input images. Using PSNR, SSIM, IFC, and NQM as evaluation criteria, the new strategy outperformed earlier techniques while maintaining simplicity and low processing resources. The model was trained on an Intel processor with a 3.10 GHz clock speed and 16 GB of RAM. The researchers observed that the model's performance could be enhanced if a more intricate model along with a much bigger and diverse dataset is used. This innovative study demonstrated the possibility of deep learning to solve SISR, a previously difficult computer vision task.

In subsequent work, Dong et al. (2016) enhanced the computational capacity of their previously proposed SRCNN network. They argued that, even though their previously suggested SRCNN outperformed hand-crafted methods, its high computational cost precluded its use in situations demanding real-time performance with limited computer capacity. In response, they suggested accelerating the earlier SRCNN model. In contrast to the SRCNN, which used pre-upsampling, the proposed enhancement used post-upsampling, which shifted the computationally costly upsampling layer to the network's last layers. The proposed network substituted a learnable deconvolution network for the initial bicubic interpolation method used for upsampling, which reduced the networks' size and learned directly from the LR input. To constrain mapping to a low-dimensional feature space, mapping layers were reduced at the beginning, and expanded at the end, therefore minimizing the cost of nonlinear mapping. They divided the mapping into numerous layers with lower filter sizes. Consequently, the completed structure resembled an hourglass, with narrow middle and thick ends. This model almost quadrupled the network's performance, had superior reconstruction quality, and achieved CPU performance in real-time. The resulting model, named FSRCNN, beats the SCN, SRF, SRCNN, and SRCNN-ex in terms of computational speed and restoration quality.

Tai et al. (2017), employ residual networks in both local and global ways to mitigate the problems associated with training very deep networks while increasing depth and utilizing recursive learning to manage model parameters. They suggested the DRRN, an extremely deep CNN model with up to 52 convolutional layers that aims for dense yet condensed networks. According to their comprehensive benchmark study, the suggested DRRN outperformed conventional CNN models for SISR while employing fewer parameters.

The authors observed that, although obtaining outstanding performance, very deep networks required huge parameters. Additionally, large models require more storage space than small models, which limits their use for mobile systems. Therefore, the innovative DRRN was suggested to efficiently design very deep network architectures to help address these challenges. In comparison to VSDR, DRCN, and RED 30, the DRRN has 2x, 6x, and 14x fewer parameters, respectively.

Lim et al. (2017) introduced the EDSR, a popular benchmark used in comparative studies for the SISR task. The EDSR architecture was developed with the goal of cutting down on the model's parameter size while maintaining a high degree of performance. The authors achieved this by using a mix of residual blocks and dense connections, both of which enable an effective flow of information throughout the network. In addition, the authors suggested using global residual learning, a technique that contributed to a further reduction in the total number of model parameters. The residual block of the EDSR has its batch normalization layers and ReLU activation function at the last layer removed as compared to the standard ResNet architecture. Lim et al. (2017) argued that the batch normalization layer causes a loss of scale information and reduces the range flexibility of the activations. Experimental results from the EDSR design confirms that the removal of the layers contributed to the memory efficiency and the overall performance of the SR model. Additionally, the authors introduced a novel upsampling process named "sub-pixel convolution." which enhanced the resolution of the generated output images and removed checkerboard effects as opposed to the transposed convolution operation. The authors investigated the proposed EDSR model on different benchmark datasets for SISR. Using the SSIM and PSNR as quantitative measures demonstrates higher performance compared to existing SISR models.

3   Methodology

In Fig. 1, we show the design of our variational autoencoder, including information on the encoder and decoder.

3.1   Encoder Model

The encoder used in this work uses a ResNet architecture inspired by that used in EDSR (Lim et al. 2017) as a feature extractor in the initial layers. As opposed to the standard ResNet, the batch normalization layers are removed along with the ReLU activation at the final layer. Experimental results from the EDSR design showed that removing the batch normalization layers reduces memory consumption while improving the overall model's performance. Inspired by the EDSR's impressive results in the context of information

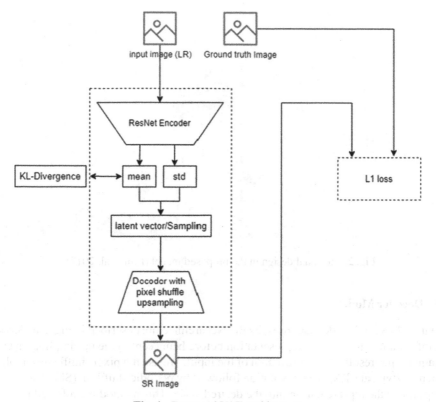

Fig. 1. Proposed VAE architecture.

preservation and memory and computational efficiency, this study employs the EDSR's ResNet design as a feature extractor in the encoder to extract high-level information in the LR images. In the proposed study, the residual blocks are stacked together in the initial layers of the encoder.

The later layers of the encoder flatten the output from the residual network and output the standard deviation (σ) and mean (μ) which are the outputs of a dense layer. To sample into the latent representation, a random Gaussian noise is generated with the standard deviation and mean from the encoder output and combined with the reparameterization trick in Eq. 1 to generate the latent representation which is used by the decoder to perform the SR reconstruction.

$$z = \mu + \sigma \odot \epsilon \tag{1}$$

Figure 2 shows the residual block used in the proposed encoder.

We found that using this network design improves the performance of the network and allows faster training on limited resources as alluded to by Lim et al. (2017).

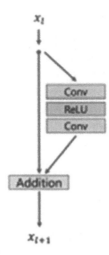

Fig. 2. Residual design in the proposed model (Lim et al. 2017)

3.2 Decoder Model

The initial layer of the decoder uses the ResNet architecture described in Fig. 2 to decode the information in the latent representation before being sent to the upsampling layer to obtain a super-resolved reconstruction of the input. We use the pixel shuffle upsampling which is a typical LR space convolution followed by periodic shuffling (Shi et al. 2016.) to upsample the representation into the desired scale. This method is preferred because it yields higher-quality representations than conventional transposed convolution, which has a narrower receptive field and processes less contextual information, yielding less accurate predictions which are also susceptible to checkerboard artifacts.

We perform the pixel shuffling operation shown in Eq. 2 and rescale the output from the initially rescaled [0, 1] range back to the [0,2 55] range.

$$I^{SR} = P\left(W \cdot f^{L-1}\left(I^{LR}\right) + b\right) \tag{2}$$

where P is the periodic shuffling operation (Shi et al. 2016) b and W are the kernel and bias respectively. The function $f^{(L-1)}$ represent the output from the previous network layer.

3.3 Loss Function

As in a standard variational autoencoder we use the KL divergence loss (see Eq. 3), a distance measure between probability distributions is used as a regularization parameter to enforce that the distribution of z (represented by mu and sigma) is close to N (0, 1). Minimizing the KL-divergence allows the network to optimize μ and σ such that they are remarkably close to that of the target distribution. Constraining the mean and standard deviation to 0 and 1 respectively ensures that the decoder always produces realistic images even when the encoder produces random noise. Minimizing just the

KL-divergence places the encodings at random toward the latent space's center without much consideration of the similarity of the nearby encodings. This makes it hard for the decoder to reconstruct any useful output because the random placement makes it difficult to extract any meaning. To construct realistic images that have regard to the similarity between nearby encodings while being centered around the latent space, we use the KL-divergence together with the L1 loss which is demonstrated by Lim et al. (2017) and Zhao et al. (2015) to perform better for the SISR task.

$$\sum_{i=1}^{n} \mu_i^2 + \sigma_i^2 - \log(\sigma_i) - 1 \tag{3}$$

4 Experiments

4.1 Training Dataset

In this study, the proposed model was trained on the DIV2K dataset. The dataset contains 100 and 800 h images for valuation and training respectively. The dataset is augmented using random flips to the left and right, random cropping, and a random rotation of images by 90°. A bicubic function with a scale factor of two was used to down-sample the ground truth HR images to obtain the LR images.

4.2 Hardware

Our proposed architecture is trained on an Intel(R) Xeon(R) Gold 6230R CPU @ 2.10 GHz running a Linux-4.15.0-194-generic-x86_64-with-glibc2.27 ubuntu server with 2 T V100S-PCIE-32 GB GPUs and 32 GB of RAM.

4.3 Training Details

We utilize 16 residual blocks in our encoder and decoder with a filter size of 64 as used in the EDSR model. Table 1 gives a summary of the specification of the entire model.

Table 1. Summary of model specification.

Option	Value
# Residual blocks in the encoder	16
# Filters	64
# Parameters	1.4M
#Loss function	L1 + Kullback-Liebler

We use the Adam optimizer to train our model using a piecewise constant decay learning rate scheduler with learning rates ranging from 1e−4 to 5e−5. The model is trained on the div2k dataset for 100 epochs with a batch size of 16 and 200 steps per epoch. Table 2 summarizes the hyperparameters utilized for experimental training.

Table 2. Hyperparameters used in model training.

Hyperparameter	Value
Epoch	100
Steps per epoch	200
Optimizer	Adam
Learning rate	[1e−4, 5e−5]
Batch size	16

4.4 Experimental Results

Our model is evaluated on the validation set of the Div2k dataset. The model is assessed using the SSIM and PSNR measures as well as the perceptual quality as observed by an individual observer. To investigate the effect of the latent dimension on the quality of the reconstruction, we experimented with various latent dimensions of 128, 1024, and 2048 using the same model architecture and hyperparameters as described in Table 2. Table 3 compares the model's performance on the different latent dimensions in terms of training time, PSNR, SSIM, and total loss. Figures 3 and 4 show the graphical comparison of the PSNR and loss values respectively.

Table 3. Performance comparison of the proposed model on different latent dimensions

Latent Dimension	Training Time	PSNR	SSIM	Loss
128	1 h 4 m 30 s	32.33	0.88	6.7
1024	1 h 30 m 10 s	33.63	0.89	5.34
2048	2 h 1 m 41 s	33.86	0.90	5.44

We observe that the PSNR and SSIM values increase as the size of the latent dimension increases. It can also be seen that the training time increases as the latent dimension increases. From this observation, it is sufficient to conclude that the latent dimension in the variational autoencoder has a direct influence on the level of detail that is learned and further reconstructed in the decoder.

It is also observed that the model achieves better perceptual quality with increasing latent dimensions. Figures 5, 6 and 7 show the LR and HR images together with the reconstructed SR images of our model on the various latent dimensions.

4.5 Comparative Evaluation

We demonstrate the performance of our model by comparing to state-of-the-art approaches. We compare with the traditional bicubic interpolation and other well-performing DL-based methods. Table 4 shows a comparison of our model to 4 other models on the Div2k dataset in terms of PSNR and SSIM.

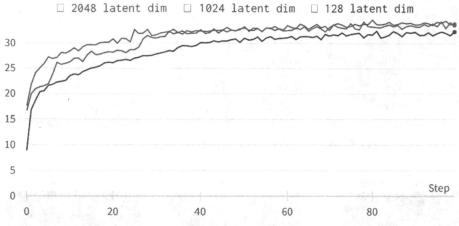

Fig. 3. Comparison of PSNR values on different latent dimensions

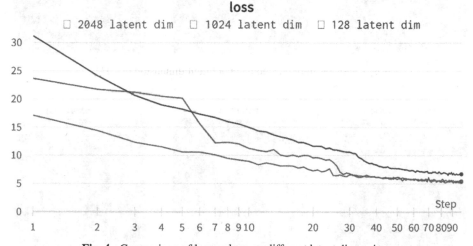

Fig. 4. Comparison of loss values on different latent dimensions

From the results obtained from the experiments on the proposed model and the comparison to other state-of-the-art approaches, we conclude that:

1. The choice of encoder and decoder architectures affects the performance of variational autoencoders to the Super-resolution task. The experiments show the successful transfer of some techniques used in other super-resolution approaches such as the ResNet design of the EDSR (Lim et al. 2017) for the encoder and decoder design of the proposed model as well as the pixel shuffle upsampling used in the last layers of the decoder model.
2. The size of the latent dimension in a variational autoencoder affects the quality of the reconstruction regarding the super-resolution task. From the experiments, it can be

100   J. K. Appati et al.

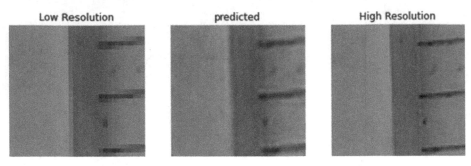

Fig. 5. SR result of our model on a latent dimension of 128.

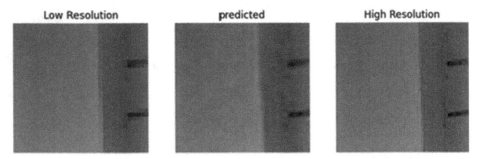

Fig. 6. SR result of our model on a latent dimension of 1024.

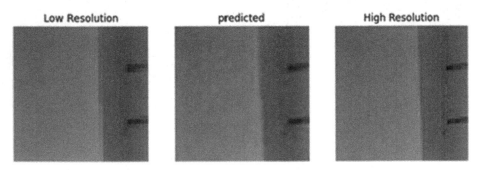

Fig. 7. SR result of our model on a latent dimension of 2048.

concluded that the larger the latent dimension, the better the reconstruction. This also
sufficiently answers the second research question and fulfills the second objective of
this research.

3. The proposed model strongly competes with some state-of-the-art models on benchmark evaluation metrics and datasets. In terms of perceptual quality, our model is
effective at removing checkerboard effects in the LR images. Although our model
lags behind the EDSR model, we demonstrate strongly that variational autoencoders
are efficient at the SISR task and can be further explored to achieve impressive results.

Table 4. Comparison of our model to other state-of-the-art-models

Model	PSNR	SSIM
Bicubic	31.01	0.9393
A+ (Timofte et al. 2015)	32.89	0.957
SRCNN (Dong et al. 2016)	33.05	0.958
VDSR (Kim et al. 2016)	33.66	0.963
EDSR (Lim et al. 2017)	35.03	0.9625
Ours	**33.86**	**0.903**

5   Conclusion

In this paper, the combination of various techniques that have shown promising results for the single image super-resolution task were used in a variational autoencoder for the solution task. A customized vanilla variational autoencoder with decoder and encoder models based on the residual network design of the EDSR model is proposed and evaluated on the Div2K super-resolution dataset with standard evaluation metrics. Results from experiments show that the proposed model attains impressive results and has comparative results to existing models and trains fast even on limited infrastructure.

With the results obtained from this paper, we allude to the assertion by Chira et al. (2022) that the generative abilities of variational autoencoders can be further explored for the SR task and have the potential of even outperforming state-of-the-art in both qualitative and quantitative measures with shorter training times on limited computational resources. Given enough time and resources, other variants of variational autoencoders that have shown significant results in other computer vision tasks can be explored together with some of the techniques used in this paper to further exploit the generative powers of variational autoencoders for the SISR task.

Abbreviation

CNNs: Convolutional Neural Networks	NQM: Noise Quality Measure
CPU: Central Processing Unit	ProSRGAN: Progressive Super-Resolution Generative Adversarial Network
DL: Deep Learning	PSNR: Peak Signal to Noise Ratio
DRCN: Deep Recursive Convolution Network	RAM: Random Access Memory
DRRN: Deep Recursive Residual Network	RED: Residual Encoder-Decoder
EDSR: Enhanced Deep Residual Network	SCN: Sparse Coding Based Network
ESRGAN: Enhanced Super-Resolution Generative Adversarial Network	SR: Super Resolution

(*continued*)

(*continued*)

FSRCNN: Faster Super-Resolution Convolutional Neural Network	SRCNN: Super-Resolution Convolutional Neural Network
GANs: Generative Adversarial Networks	SRF: Super-Resolution Forest
GPUs: Graphics Processing Units	SRGAN: Super-Resolution Generative Adversarial Network
HR: high-resolution	SSIM: Structural Similarity Index Measure
IFC: Information Fidelity Criterion	SSIR: Single Image Super-Resolution
KL: Kullback-Liebler	VAEs: Variational Autoencoders
LR: low resolution	VDVAE: Very Deep Variational Autoencoder
NN: Neural Networks	

References

Chen, X., Mishra, N., Rohaninejad, M., Abbeel, P.: PixelSNAIL: an improved autoregressive generative model. In: International Conference on Machine Learning, pp. 864–872. PMLR, July 2018

Chira, D., Haralampiev, I., Winther, O., Dittadi, A., Liévin, V.: Image super-resolution with deep variational autoencoders. arXiv. https://doi.org/10.48550/arXiv.2203.09445 (2022)

Dodge, S., Karam, L.: Understanding how image quality affects deep neural networks. In: 2016 8th International Conference on Quality of Multimedia Experience, QoMEX 2016 (2016). https://doi.org/10.1109/QoMEX.2016.7498955

Goodfellow, I.J., Mirza, M., Xu, B., Ozair, S., Courville, A., Bengio, Y.: Generative adversarial networks. arXiv. https://doi.org/10.48550/arXiv.1406.2661 (2014)

Goodfellow, I., Bengio, Y., Courville, A.: Deep Learning. MIT Press, Cambridge (2016)

He, K., Zhang, X., Ren, S., Sun, J.: Deep residual learning for image recognition. http://arxiv.org/abs/1512.03385 (2015)

Irani, M., Peleg, S.: Improving resolution by image registration. CVGIP Graph. Models Image Process. **53**(3), 231–239 (1991). https://doi.org/10.1016/1049-9652(91)90045-1

Keys, R.: Cubic convolution interpolation for digital image processing. IEEE Trans. Acoust. Speech Signal Process. **29**(6), 1153–1160 (1981). https://doi.org/10.1109/tassp.1981.1163711

Kingma, D.P., Welling, M.: Auto-encoding variational bayes. arXiv. https://doi.org/10.48550/arXiv.1312.6114 (2013)

Krizhevsky, A., Sutskever, I., Hinton, G.: ImageNet classification with deep convolutional neural networks. Commun. ACM **60**(6), 84–90 (2017). https://doi.org/10.1145/3065386

Lim, B., Son, S., Kim, H., Nah, S., Lee, K.M.: Enhanced deep residual networks forsingle image super-resolution. In: 2017 IEEE Conference on Computer Vision and Pattern Recognition Workshops (CVPRW), pp. 1132–1140 (2017)

Sarode, R., Varpe, S., Kolte, O., Ragha, L.: Image super resolution using enhanced super resolution generative adversarial network. In: ITM Web Conferences, vol. 44, pp. 03054 (2022). https://doi.org/10.1051/itmconf/20224403054

Shi, W., et al.: SR with sub-pixel. Comput. Vis. Pattern Recognit., 1874–1883 (2016)

Tai, Y., Yang, J., Liu, X.: Image super-resolution via deep recursive residual network. In: Proceedings of the IEEE Conference on Computer Vision and Pattern Recognition, pp. 3147–3155 (2017)

Timofte, R., De Smet, V., Van Gool, L.: A+: Adjusted anchored neighborhood regression for fast super-resolution. In: Cremers, D., Reid, I., Saito, H., Yang, MH. (eds) ACCV 2014. LNCS, vol. 9006, pp. 111–126. Springer, Cham (2015). https://doi.org/10.1007/978-3-319-16817-3_8

Uria, B., Côté, M.A., Gregor, K., Murray, I., Larochelle, H.: Neural autoregressive distribution estimation. J. Mach. Learn. Res. **17**(1), 7184–7220 (2016)

Wang, Z., Chen, J., Hoi, S.: Deep learning for image super-resolution: a survey. IEEE Trans. Pattern Anal. Mach. Intell. **43**(10), 3365–3387 (2021). https://doi.org/10.1109/tpami.2020.2982166

Zhao, H., Gallo, O., Frosio, I., Kautz, J.: Loss functions for neural networks for image processing. arXiv preprint arXiv:1511.08861 (2015)

Hybrid HAN Model to Investigate Depression from Twitter Posts

Salma Akter Asma[1] (ID), Nazneen Akhter[1(✉)] (ID), Mehenaz Afrin[1(✉)] (ID),
Sadik Hasan[1(✉)] (ID), Md. Saide Mia[1(✉)] (ID), and K. M. Akkas Ali[2(✉)] (ID)

[1] Bangladesh University of Professionals, R9Q5+Q7G, Dhaka, Bangladesh
nazneenmow66@gmail.com, agroti202@gmail.com,
sadik.h.emon@gmail.com, sayed.bup@gmail.com
[2] Janagirnagar University, Savar 1342, Bangladesh
akkas@juniv.edu

Abstract. Depression is currently one of the most concerning complications in the world. With the availability of social networking sites, human openly express their feelings which is used for depression detection. Manually it is tough to work with millions of mental disorders peoples. To overcome the time complexity of the manual system, researchers are adopting automation technologies to detect depression by analyzing social media posts. However, the majority of existing ML models do not provide any explanation for the model's behavior. Furthermore, ML models do not learn high-level features incrementally, resulting in lower accuracy when detecting depression from a longer sequence of data. So, a hybrid Hierarchical Attention Network (HAN) is implemented into this model. The hybrid HAN model is intrigued with BiGRU at the word phase and BiLSTM at the sentence phase attention mechanism. Hybrid HAN model provides an explanation of the model and improves contextual dependency in the long-term sequence, with an accuracy of 97.09%.

Keywords: Depression Detection · Deep Learning · Hybrid Hierarchical Attention Network · Natural Language Processing

1 Introduction

Depression is a temperamental disease in which intent is a diligence intuition of boredom or ruin of interest. Nowadays, it is one of the most alarming complications in the world. Usually, it affects not only professional life but also family and other relationships [1]. Every year, nearly 0.8 million adolescent depressed adults decide to suicide [2]. More than half of the people in the world have been suffering from depression lightly [3]. Actually, depression is dealt with smoothly when it is recognized promptly [4]. In response to the WHO, in 2020, mental illness affected nearly two hundred million individuals worldwide [5].

A later study showed that almost 70% of young people utilize social media for more than five hours [6]. Internet sharing is a global imperative and social media networks in communication technology fascinate people to interact with each other from anywhere in the world [7].

Social media is used to help people share their thoughts, feelings, desires, achievements, and more [8]. Those can be considered as post-diagnosis analysis and can be helped to investigate their depression.

This work follows the following format: The background study is covered in Sect. 2, and the depression detection model contains in Sect. 3. The fourth section describes the comparisons and performance analysis of the model, while the fifth section addresses the conclusion and future development.

2 Background Study

The ease of using social media sites is accelerating researchers to work in this field. Researchers are proposing different types of depression detection models based on ML and DL. In the year 2020, Mustafa et al. presented an ML model that can predict depression from Twitter posts [9]. They focused on some words and assigned weights on them. Accordingly, weighted words were utilized with 14 mental qualities with Etymological Request and LIWC to categorize those words into their own classes of moods.

Ma et al. proposed a model for depression detection with the voting method [10]. The author evaluated the result with MNB, GB, RF, and Voting classifiers. They applied one hot encoding to give focus on words. They achieved 85.09% of accuracy to detect depression. LDA, and LSA were utilized Koltai et al. [11].

Gaind et al. (2019) divided the social media information that they investigated into six categories, including joy, pity, fear, indignation, shock, and disgusting [12]. Using the SMO classifier, they were able to attain 85.4% accuracy and 91.7% precision for J48.

A model using Anaphora resolution was put up by Wongkoblap et al. (2021) to identify depression from tweets [5]. In order to specifically categorize depression, they introduced Multiple Instance Learning in this research and created a pair of models, MIL-SocNet and Multiple Instance Learning utilizing an Anaphoric Approach for Social Network. Their anaphoric model's accuracy was 92%.

Amanat et al. suggested a depression detection modeL integrating LSTM and RNN [13]. They pre-processed the data by using transformations and normalization. Uddin et al. suggested an LSTM-based RNN algorithm to identify the symptoms of depression from youths' personal text-based queries. On a public Norwegian website: ung.no for young people of Norway [14]. The approached dataset achieved 98% accuracy using 11,807 texts and the second dataset achieved 99% accuracy using 21,807 texts. With the LSTM-RNN Gaafar et al. achieved 96.50% to 85.69% accuracy with images [15]. They identify hidden trends and characteristics in datasets of text as well as pictures. Kim et al. (2022) demonstrated an investigation of depression making use of the PHQ-9 and NLP based on social media text [16]. To categorize the user's depression, they applied a logistic regression model. The overall accuracy was 68.3%, which was higher than the baseline depression classifier's 53.3% average accuracy from employing Y/N sentence classifiers.

Naseem et al. investigate 4 depression classes minimal, Mild, Moderate, and Severe by utilizing attention-based BiLSTM [17]. They used traditional methods (SVM, RF, and MLP), Graph based methods (TextGCN, TensorGCN) with cross-entropy, and deep learning-based methods (CNN, RNN, DepressionNet). They achieved an accuracy of 85% and D2 of 95% for their model. Zogan et al. proposed a hierarchical Attention Network (HAN) in user tweets to find the relevant information from text data [18]. BiGRU neural network is used in both sentence and word encoder models to implement the HAN model.

SenseMood architecture was proposed by Lin et al. [19]. The system Sense Mood can be utilized both online and offline. For offline detection, the textual features are trained with the BERT model. The Convolutional Neural Network model is applied to visual features. These two models are merged together utilizing a low-rank tensor. The model's accuracy and F1 score were both 88.39% and 93.59%, correspondingly.

Researchers are adopting automation technologies such as ML, DL, and hybrid methods to detect depressed users from social media posts. However, most of them neglected to observe the significance of contextual analysis in lengthier linguistic sequences. On the basis of a single tweet, it might be nearly impossible to ascertain the user's mental condition. As a result, one month of tweets has been considered to analyze the depressed user in this work. To detect depressed users by contextually analyzing longer sequences of text data, a hybrid HAN model is suggested in this study.

3   Depression Detection Model

A Hybrid HAN model is put forth that intrigued with BiGRU in the word phase and BiLSTM in the sentence phase attention mechanism. The suggested model is a two-level attention system that uses words as well as sentences in a hierarchical fashion. Figure 1 depicts the general design of our suggested model. It composed of several layers depicted as Word embedding layer, BiGRU layer, TimeDistributed Dense layer, Attention Context layer, Sentence Input layer, BiLSTM layer, Dense Layer, etc. Different components of the architecture are described in the following paragraphs.

3.1   BiGRU Layer

It is an extended version of RNN which comprises an Update gate and a Reset gate [20].

$$z_t = \sigma(w_t * (h_{t-1}, \; x_t)) \tag{1}$$

Here, z_t is the update gate, which controls the extent to which data passes across the previous state to the next.

$$r_t = \sigma(w_r * (h_{t-1}, \; x_t)) \tag{2}$$

Here, r_t is the reset gate that is in charge of deciding which information will not pass in the following state.

$$\overrightarrow{h_t^w} = \overrightarrow{GRU} \, (\overrightarrow{h_{t-1}}, x_t) \tag{3}$$

$$h_t^{w\leftarrow} = GRU^{\leftarrow} \, (h_{t-1}^{\leftarrow}, x_t) \tag{4}$$

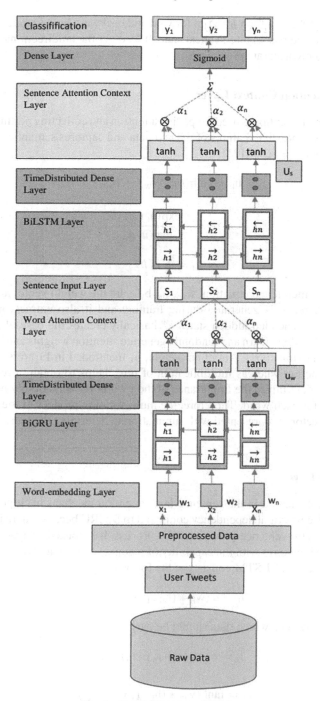

Fig. 1. Hybrid HAN Architecture

The fusion $(h_t^w = [\overrightarrow{h_t^w} \oplus h_t^{w \leftarrow}])$ of forward and backward hidden states generalize a common weight that passes into the next state. It resolves the unidirectional GRU issues by generating more accurate predictions.

3.2 Word Attention Context Layer

The mechanism of attention is used for paying attention and collecting pertinent information. It analyzes contextual features of the text data and captures semantic and syntactic information from a sentence [21].

$$u_t = \tanh((h_t^w W_{ww} + b_{ww}) \tag{5}$$

$$\alpha_t = \frac{\exp(u_t u_w)}{\Sigma_j \exp(u_t u_w)} \tag{6}$$

$$c_i = \Sigma \, \alpha_t \, h_t^w \tag{7}$$

In the above-mentioned equation, w_{ww} and b_{ww} are the weight vector and bias value of the input. Here, u_t is generated during training and it also extracts global words for all words. The encoded hidden states h_t^w basically creates the base of the attention model. Also, it lets them learn and randomly generate attention weights and biases during training for context analysis. Attention weight, α_t mentioned in Eq. 6 is generated by the softmax function whereas multiplication of $u_t u_w$ helps to obtain a feature vector. Additionally, it determines the importance of the j word. u_w is randomly produced as a context vector for each word. Therefore, the multiplication product of the hidden state and attention vector is represented with summation to determine the word-level vector, c_i.

3.3 BiLSTM Layer

BiLSTM is implemented in the sentence-level attention mechanism. It performs more accurately with contextual dependency compared to BiGRU because there is more long-term sequence in the sentence-phase attention process. It enables the capacity to remember data for the long term with the capability of reading, writing, and deleting data from the memory layer [22]. LSTM consists of the following gates.

$$f_t = \sigma(w_{ft} * (h_{tf-1}, x_{tf}) + b_{ff}) \tag{8}$$

Forget gate, f_t specifies which data should be purged.

$$i_t = \sigma(w_{ii} * (h_{ti-1}, x_{ti}) + b_{ii}) \tag{9}$$

$$c_i = \tanh(w_{ci} * (h_{ti-1}, x_t) + b_{ci}) \tag{10}$$

The input gate determines which information will be stored for a longer period that operates with two functions. The sigmoid function, i_t states which value should be updated. On the contrary, the function c_i determines which values should be added to the LSTM memory.

$$O_t = \sigma \left(WO_o * (h_{to-1}, X_t) + b_{ou} \right) \tag{11}$$

Output gate, O_t states the part of LSTM memory that has more effect on predicting the output.

BiLSTM works on both the forward and reverse direction of data, which improves analyzing semantic and syntactic information of long-term sequence.

$$\overrightarrow{h_t^w} = \overrightarrow{LSTM} (\overrightarrow{h_{t-1}}, x_t) \tag{12}$$

$$h_t^{w\leftarrow} = \overleftarrow{LSTM} (h_{t-1}^{\leftarrow}, x_t) \tag{13}$$

$$BiLSTM = [\,\overrightarrow{h_t^w} \oplus h_t^{w\leftarrow}] \tag{14}$$

BiLSTM function conveys relevant information to the next layer of the model.

3.4 Sentence Attention Context Layer

To understand the context behind the sentence representations of tweets, again attention mechanism is applied in this layer [23]. BiLSTM has been applied in this layer to improve extracting semantic and syntactic information of long-term sequences.

$$U_i = \tanh((h_i^S s_s + b_s) \tag{15}$$

$$\alpha_t = \frac{\exp(u_i\,u_s)}{\Sigma_j \exp(u_i\,u_s)} \tag{16}$$

$$s_i = \Sigma_i\,\alpha_i\,h_f \tag{17}$$

In the above-mentioned equation, s_s is the randomly initialized sentence vector, and b_s is the bias matrix. Here, u_i extracts the global sentence vector for sentences. The multiplication of $u_i\,u_s$ indicates how important the sentence in the tweet is. Therefore, a product of the attention vector and the hidden state of the tweet is represented with summation to determine sentence vector s_i, of the end-user post.

3.5 Overall Architecture of Depression Detection Model

In the suggested model, word embedding layer is implemented on the first layer. It is a dense vector representation of words according to semantic meaning. In this layer, the parameters embedding dim, vocab size, input length, and trainable are respectively set to 200, 100000, 14, and False. Weights are fed into this layer which is generated using the 'GloVe Twitter 200D' model is applied for pre-training the model. GloVe Twitter library is particularly deployed for specific words that are frequently used in tweet posts. Embedding dim and vocab size respectively denote the dimension of the vector of every word and the size of the vocabulary. Input length represents the maximum length of words that can be trained. Trainable is set to false to overcome the drawbacks of fine-tuning so that the model can perform better even with new data.

Sequentially, BiGRU is implemented on the second layer. L2 regularizer is used as a kernel regularizer in the BiGRU layer. Additionally, the L2 regularizer is tuned to the $1e-6$ regularization parameter to prevent overfitting and minimize the loss function of gradient descent. The hidden layer, allow the model to capture correlations between the current input words and the words preceding and following them. This results in more accurate predictions that consider the context of the input. The hidden layer is set to 150, which is chosen based on the complexity. Return sequence is set to "True," allowing the model to return a sequence of predicted words rather than simply the final prediction.

TimeDistributed dense layer is applied on the third layer. The amount of samples processed during each training iteration is determined by the batch size in this layer and with a value of 210. The attention mechanism function is implemented on the fourth layer with a dropout parameter of 0.8. The Word encoder model is implemented with an input function and attention context layer. It maps the input of words with an attention mechanism to decide which word vector carries more importance to determine the prediction.

The first layer of the sentence-level attention mechanism is the input function, which is applied with values of 50 and 50 in the max sentence num and max sentence len parameters. The input data in the model is controlled by two parameters, max sentence num, and max sentence len. The first parameter determines the maximum number of sentences that can be fed from each user's data into the model, while the second parameter specifies the maximum number of words in each sentence that can be inputted. The second layer contains the TimeDistributed layer with the word encoder model mapping function. The aforementioned layer has created a hierarchy among words and sentence-phase.

BiLSTM is implemented in the third layer of the sentence-phase attention mechanism. The hidden layer is set to 180, which is chosen based on the complexity. Return sequences are set to 'True' for returning a sequence of predicted words instead of just the final prediction.

The fourth layer is implemented with a Timedistributed dense layer with a batch size of 215. The attention context layer is implemented in the fifth layer with a dropout parameter of 0.8. Units are set to 1 in the Dense layer as it is a binary classification model. Additionally, the sigmoid activation function is used in the dense layer.

Finally, the model is compiled with the 'Adam' optimizer function and the 'binary crossentropy' loss function. Batch size and epochs are set to 20 and 10 respectively in the kernel classifier. K-fold cross-validation is set to 10 to prevent overfitting and checks how the model behaves with new data.

3.6 Dataset

A large-scale dataset provided by Shen et al. is implemented in this experiment to examine the proposed approach [24]. The dataset is subcategorized into three parts, particularly depression, nondepression, and candidate. The dataset which contains only depressed users is referenced as D1 Dataset. Additionally, the dataset contains non-depressed users, and depression candidate users are referenced as D2 and D3. There were 58900 users' tweets stored in the D3 dataset.

As D3 does not determine whether users are depressed or non-depressed in this dataset, a lexicon is proposed to label the D3 dataset. The lexicon contains 811 words. If the lexicon words are found in the tweets that are marked as depressed otherwise counted as non-depressed. While the D3 dataset was labeled using the proposed lexicon, there were 1173 non-depressed and 57,727 depressed users. The labeling process is illustrated in Fig. 2.

A total of 2000 user tweets were randomly selected from three datasets, with 500 tweets each from D1 and D2 comprising both depressed and non-depressed user's, and 1000 tweets from D3, consisting of 500 tweets from depressed users and 500 tweets from non-depressed users. Subsequently, all these user tweets merged for testing and training purposes.

3.7 Lexicon

The lexicon is created considering some aspects which trigger depression such as job, financial condition, bullying, family issues, health, drug addiction, genetic condition, childhood trauma, relationship, etc. [25, 26]. The words were collected from different sources such as suicide letters, anti-depressant drugs, Vader lexicon, swear words, etc. Swear world is included in the lexicon as depressed people are prone to cursing more than normal people [27, 28]. The antidepressant name is included as they often tweet about their prescription in the tweets. Some words were collected from suicidal notes and posts that are frequently used among suicidal people. Vader lexicon contains three types of sentiment words, for labeling only negative sentiment words, were counted [29].

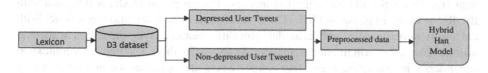

Fig. 2. D3 Labeled utilizing Lexicon

3.8 Dataset Preprocessing

For removing redundancy and noise in the dataset, preprocessing technique is adopted in this experiment. To clean the dataset, extra spaces, stopwords, punctuation, tweet mentions, URLs, and white spaces have been removed. Emoji is converted into text for context analysis. Emoji carries relevant information about user thoughts. PorterStemmer is adopted for getting the root words of every word in the text. It removes suffix morphology and inflection from words to make it easier for information retrieval. In later stages, all the words were appended together for implementing the proposed approach.

4 Performance Analysis

This section verifies the beneficial effects of the hybrid HAN model on contextual dependencies and semantic ambiguity in longer sequences. In order to evaluate the proposed hybrid HAN model, it was compared to the following algorithms: CNN, SVM, Hierarchical Attention Network (HAN), DT, and RF. Four evaluation measures used to report the outcome using the same dataset of these classifiers. This research is being carried out on a PC running Windows 10, a 64-bit OS, a Core i5 processor, and 8 GB of RAM. Two questions RQ1 and RQ2 are addressed to experimentally satisfy the model. The following analysis includes a comprehensive summary of the aforementioned research questions as well as their evaluation.

Table 1. The classifiers' effectiveness

Classifier	Accuracy	Precision	Recall	F1-Score
Hybrid HAN	97.09	96.39	96.85	96.99
HAN	96.03	95.1	96.45	96.13
CNN	80.56	77.13	79.73	74.21
SVM	76.2	75.89	74.8	79.06
RF	78.35	79.45	81.58	81.58
DT	76.95	78.29	74.6	74.9

RQ1: How does hybrid HAN enhance contextual dependencies in long sequences?
Table 1 shows that the HAN model with only BiGRU put a slightly poor performance than HAN with BiGRU and BiLSTM models. The proposed model is intrigued with BiGRU at the word phase and BiLSTM at the sentence phase attention technique. With the ability to read, write, and delete data from the memory layer, the BiLSTM mechanism enables long-term memory retention. As shown in Table 1, it improves contextual dependency in predicting users' one-month tweets and achieves the highest accuracy of 97.09%. The HAN model, on the other hand, achieves 96.09% using only BiGRU. In comparison to BiLSTM, BiGRU mechanisms have limitations when it comes to storing long-term sequences in memory. As a result, it provides a slightly subpar performance. Table 1 shows that SVM, RF, and DT achieve an accuracy of 76.2%, 78.35%, and 76.95%. Because the Machine Learning algorithm is incapable of learning long sequences incrementally, it cannot comprehend the context of the user's one-month tweet.

CNN outperforms the machine learning model but falls short of the hybrid HAN model. It has an 80.56% accuracy rate. As CNN has convolutional layers, it grasps higher-level feature information incrementally and produces better results than SVM, RF, and DT. However, CNN performs poorly in comparison to the hybrid HAN model since it cannot hold data from longer sentences and is unable to comprehend the context of tweets.

Table 2. Performance of proposed model and other author's comparative proposed model on Twitter dataset

Authors	Dataset Source	Used model	Accuracy
Zogan et al. 2020 [18]	Twitter (Public/Pu)	HAN, MDHAN	84.4, 89.5
Shen et al. 2017 [24]	Twitter(Pu)	MDL	84.8
Lin et al. 2020 [19]	Twitter(Pu)	CNN	88.4
Proposed model	Twitter(Pu)	Hyrbid HAN	97.09

Table 2 presents a comparative analysis of some qualitative research findings obtained from the same dataset. It should be taken into account that nearly all of the studies ignored the case of improving contextual dependencies in longer sequences of text data. As a result, they obtained poor performance, whereas the proposed model in this paper achieved 97.09% when considering the importance of contextual dependencies in longer sequences of text data.

RQ2: How can semantic ambiguity be resolved by the hybrid HAN to improve user tweet predictive performance?

The same words used in different sentences may have different interpretations. For example, 'When my mother scolds me, I feel suffocated'. Here, 'Suffocated' emphasizes the importance of identifying depression more than any other word. Another example is 'People start to feel suffocate when there isn't enough oxygen,' which contains the word suffocated but isn't recognized as a depressing tweet because the context is different here. The proposed model utilize a two-level hierarchical attention mechanism that uses words and sentences in a hierarchical fashion to improve this type of semantic ambiguity.

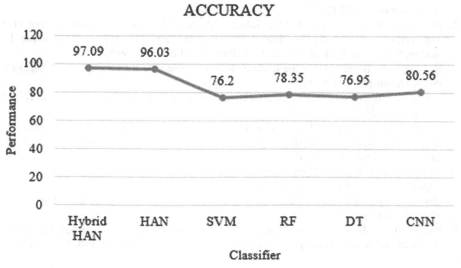

Fig. 3. The accuracy of the classifier's

To predict tweets, the word-level attention mechanism model only focuses on that individual word part to assign a weight, ignoring the context. However, at the sentence level, the attention mechanism for that word looks at its surroundings and assigns a different weight based on its context. As a result, hybrid HAN models achieve excellent results. Despite the fact that the CNN is a deep learning algorithm but did not perform well, as shown in Fig. 3. Since the hybrid HAN model acknowledges the longer textual context, it can deal with semantic ambiguity and thus provides a more accurate prediction of the user's mental state.

5 Conclusion

On social networking sites, most people openly express their thoughts, feelings, and activities, which is quite useful for determining their mental health. As part of this research, a deep learning-based hybrid approach is suggested for detecting depression in tweets. Preprocessing techniques are used for removing redundancy and noise in the dataset. This hybrid HAN model is applied with BiGRU and BiLSTM neural networks. BiLSTM works better with a longer sequence of text data compared to BiGRU. Our proposed hybrid HAN model attains the highest accuracy of 97.09%. According to the outcome analysis, hybrid HAN model performs well to detect depression from tweets. Only tweet-based analysis is suggested in this work; in future studies, users imagery, information such as age and occupation will be assessed further to analyze depression in greater detail.

References

1. Chakraborty, S., Mahdi, H.F., Ali Al-Abyadh, M.H., Pant, K., Sharma, A., Ahmadi, F.: Large-scale textual datasets and deep learning for the prediction of depressed symptoms. Comput. Intell. Neurosci. **2022** (2022)
2. Häfner, H., Maurer, K., Trendler, G., Schmidt, M., et al.: The early course of schizophrenia and depression. Eur. Arch. Psychiatry Clin. Neurosci. **255**(3), 167–173 (2005). Author, F., Author, S., Author, T.: Book title. 2nd edn. Publisher, Location (1999)
3. Kessler, R.C., Bromet, E.J.: The epidemiology of depression across cultures. Annu. Rev. Public Health **34**, 119 (2013)
4. Depression, W.: Other Common Mental Disorders: Global Health Estimates. World Health Organization 24, Geneva (2017)
5. Wongkoblap, A., Vadillo, M.A., Curcin, V., et al.: Deep learning with anaphora resolution for the detection of tweeters with depression: algorithm development and validation study. JMIR Mental Health **8**(8), 19824 (2021)
6. Organization, W.H., et al.: Comprehensive mental health action plan 2013-2020-2030 (2013). https://www.who.int/mentalhealth/actionplan2013/en
7. Halfin, A.: Depression: the benefits of early and appropriate treatment. Am. J. Manag. Care **13**(4), 92 (2007)
8. Picardi, A., et al.: A randomised controlled trial of the effectiveness of a program for early detection and treatment of depression in primary care. J. Affect. Disord. **198**, 96–101 (2016)

9. Mustafa, R.U., Ashraf, N., Ahmed, F.S., Ferzund, J., Shahzad, B., Gelbukh, A.: A multiclass depression detection in social media based on sentiment analysis. In: Latifi, S. (eds.) 17th International Conference on Information Technology–New Generations (ITNG 2020). AISC, vol. 1134, pp. 659–662. Springer, Cham (2020). https://doi.org/10.1007/978-3-030-43020-7_89

10. Ma, L., Wang, Z., Zhang, Y.: Extracting depression symptoms from social networks and web blogs via text mining. In: Cai, Z., Daescu, O., Li, M. (eds.) ISBRA 2017. LNCS, vol. 10330, pp. 325–330. Springer, Cham (2017). https://doi.org/10.1007/978-3-319-59575-7_29

11. Koltai, J., Kmetty, Z., Bozsonyi, K.: From durkheim to machine learning: finding the relevant sociological content in depression and suicide-related social media discourses. In: Rudas, T., Péli, G. (eds.) Pathways Between Social Science and Computational Social Science. CSS, pp. 237–258. Springer, Cham (2021). https://doi.org/10.1007/978-3-030-54936-7_11

12. Gaind, B., Syal, V., Padgalwar, S.: Emotion detection and analysis on social media. arXiv preprint arXiv:1901.08458 (2019)

13. Amanat, A., et al.: Deep learning for depression detection from textual data. Electronics 11(5), 676 (2022)

14. Uddin, M.Z., Dysthe, K.K., Følstad, A., Brandtzaeg, P.B.: Deep learning for prediction of depressive symptoms in a large textual dataset. Neural Comput. Appl. 34(1), 721–744 (2022)

15. Gaafar, A.S., Dahr, J.M., Hamoud, A.K.: Comparative analysis of performance of deep learning classification approach based on LSTM-RNN for textual and image datasets. Informatica 46(5) (2022)

16. Kim, N.H., Kim, J.M., Park, D.M., Ji, S.R., Kim, J.W.: Analysis of depression in social media texts through the patient health questionnaire-9 and natural language processing. Digit. Health 8, 20552076221114204 (2022)

17. Naseem, U., Dunn, A.G., Kim, J., Khushi, M.: Early identification of depression severity levels on reddit using ordinal classification. In: Proceedings of the ACM Web Conference 2022, pp. 2563–2572 (2022)

18. Zogan, H., Razzak, I., Wang, X., Jameel, S., Xu, G.: Explainable depression detection with multi-modalities using a hybrid deep learning model on social media. arXiv preprint arXiv: 2007.02847 (2020)

19. Lin, C., et al.: Sense-mood: depression detection on social media. In: Proceedings of the 2020 International Conference on Multimedia Retrieval, pp. 407–411 (2020)

20. Dey, R., Salem, F.M.: Gate-variants of gated recurrent unit (GRU) neural networks. In: 2017 IEEE 60th International Midwest Symposium on Circuits and Systems (MWSCAS), pp. 1597–1600 (2017). IEEE

21. Dhingra, B., Liu, H., Yang, Z., Cohen, W.W., Salakhutdinov, R.: Gated-attention readers for text comprehension. arXiv preprint arXiv:1606.01549 (2016)

22. Siami-Namini, S., Tavakoli, N., Namin, A.S.: The performance of LSTM and BiLSTM in forecasting time series. In: 2019 IEEE International Conference on Big Data (Big Data), pp. 3285–3292 (2019). IEEE

23. Yang, Z., Yang, D., Dyer, C., He, X., Smola, A., Hovy, E.: Hierarchical attention networks for document classification. In: Proceedings of the 2016 Conference of the North American Chapter of the Association for Computational Linguistics: Human Language Technologies, pp. 1480–1489 (2016)

24. Shen, G., et al.: Depression detection via harvesting social media: a multimodal dictionary learning solution. In: IJCAI, pp. 3838–3844 (2017)

25. Khalsa, S.-R., McCarthy, K.S., Sharpless, B.A., Barrett, M.S., Barber, J.P.: Beliefs about the causes of depression and treatment preferences. J. Clin. Psychol. 67(6), 539–549 (2011)

26. Kaltiala-Heino, R., Fröjd, S.: Correlation between bullying and clinical depression in adolescent patients. Adolesc. Health Med. Ther. 2, 37 (2011)

27. Sadasivuni, S.T., Zhang, Y.: Analyzing the bad-words in tweets of Twitter users to discover the mental health happiness index and feel-good-factors. In: 2021 International Conference on Data Mining Workshops (ICDMW), pp. 882–888 (2021). IEEE

28. Rahul, H.: Assessment of the depression-level effectiveness of the curse words in young adults in private co-educational pharmaceutical institutions in Pune university pharmaceutical institutions living with poor sanitation, India: a pre-planned, causal pathway-based analysis. India: A Pre-planned, Causal-Pathway-Based Analysis, pp. 6–10 (2018)

29. Hutto, C., Gilbert, E.: VADER: a parsimonious rule-based model for sentiment analysis of social media text. In: Proceedings of the International AAAI Conference on Web and Social Media, vol. 8, pp. 216–225 (2014)

Transfer Learning-Based Encoder-Decoder Model for Skin Lesion Segmentation

Justice Kwame Appati⬤, Leonard Mensah Boante⬤, Ebenezer Owusu⁽✉⁾⬤, and Silas Kwabla Gah⬤

Department of Computer Science, University of Ghana, Accra, Ghana
{jkappati,ebeowusu}@ug.edu.gh, {lboante,skgah001}@st.ug.edu.gh

Abstract. Skin cancer is a highly dangerous form of cancer that affects numerous countries. Studies have shown that the timely detection of melanoma or skin cancer can lead to improved survival rates. Patients diagnosed with melanoma cancer at an early stage have a 90% chance of survival and tend to respond well to treatment. With this understanding, this study seeks to develop various architectures for the early detection of melanoma. Transfer learning, an approach that has garnered significant attention among researchers in solving computer vision problems, was employed in this project. In this study the U-Net architecture's encoder was modified by replacine it with a pre-trained model to enhance its performance. The performance of the proposed segmentation model was evaluated using the ISIC-2018 dataset. The model recorded a dice coefficient score of 90.7% which is a 4.7% improvement on U-Net model (86%) for segmenting skin lesions. The model's performance was further evaluated using other metrics such as recall (91.86%) and precision (91.13%). Subsequent analysis was conducted to determine the best hyper-parameters that provide the highest degree of performance when segmenting skin lesions. The results revealed that using the Efficient-Net pre-trained model as the encoder, PReLu activation function, and Tversky loss function yielded better performance in segmenting skin lesions.

Keywords: Encoder · Transfer Learning · Segmentation · U-Net architecture

1 Introduction

Globally, skin cancer has remained the most prevalent disease, with melanoma being the deadliest form, accounting for about 75% of all skin lesion cases. Early detection of cancerous cells is critical to improving patient survival rates. Therefore, several techniques have been developed for the early diagnosis of skin cancer. Dermatologists typically conduct skin biopsies to determine if a patient has cancer. While experienced dermatologists have made significant contributions to the detection and treatment of many cancer patients, there have been instances of misdiagnosis, resulting in false positives or true negatives [1]. Because of this, the application of machine learning and image processing has also been implemented to aid dermatologists in their diagnosis. Image processing methods have been adopted in computer-aided diagnosis in extracting relevant descriptors from skin lesion images. This descriptors are then used to predict the presence of

J. Abawajy et al. (Eds.): ICICCT 2023, CCIS 1841, pp. 117–128, 2023.
https://doi.org/10.1007/978-3-031-43838-7_9

skin cancer when new images are introduced to the model. However, this approach has limitations, as it requires expert selection of relevant features, which can be prone to errors. It also demands a high level of expertise and consensus among practitioners or teams on the features and methods employed in feature extraction [2]. To address the challenge of skin lesion segmentation, machine learning techniques were employed to learn extracted features and enhance the model's ability to work with unseen data. A large dataset is collected, preprocessed, and divided into train and test sets. The features are then extracted and subjected to dimensionality reduction and feature engineering techniques to develop a robust model. By training the model with the preprocessed data, it learns to recognize the relevant features and achieves excellent performance on unseen datasets [3].

Moreover, researchers also discovered some lapses in these approaches since the selection of the features during the feature engineering process was still prone to human errors [4]. The advent of deep learning has enabled machines to learn autonomously without explicit instructions on feature extraction. This remarkable feat has empowered deep learning models to learn from raw data, thereby determining the most appropriate features to describe the problem at hand. Consequently, the efficacy of deep learning models is highly dependent on the availability of massive datasets for training. Furthermore, the successful construction of a robust architecture significantly enhances a model's accuracy on a given dataset. Notably, CNNs have gained considerable traction in the field of computer vision, particularly for disease classification and other related image-processing tasks [5]. Disease classification models have shown remarkable success in detecting various diseases. However, certain diseases such as cancer present a "needle in a haystack" challenge, and it is highly recommended by researchers to implement semantic segmentation to classify images at the pixel level for accurate detection. Different architectures have been utilized to perform semantic segmentation on healthcare images.

Currently, U-Net which comprises a decoder and an encoder for image segmentation at the pixel level is judged the state-of-the-art segmentation architecture. However, training neural networks from scratch typically requires a vast amount of data that may not be readily available. Therefore, transfer learning is commonly used as a technique for weight initialization using pre-trained model's weights. As the encoder part of the U-Net architecture is a simple CNN, this study seeks to enhance the model's accuracy and training time by replacing it with a pre-trained model.

1.1 Study Objectives

The central focus of this study is to propose an efficient encoder architecture based on transfer learning for the classical U-Net architecture. Also we seek to conduct an experiment to identify the optimal combination of the objective function (loss function) and activation function for an efficient skin lesion segmentation. Finally, we seek to provide a comprehensive analysis of the proposed encoder-decoder model.

This study is structured as follows: in Sect. 2, we present a detailed review of related work on skin cancer localization and diagnosis, including latest techniques and their limitations. This section will provide a extensive understanding of the current research landscape and highlight the gaps and challenges that the proposed model seeks to address.

In Sect. 3, we describe the proposed methodology and the experimental setup. We provide a detailed overview of the architecture of the encoder-decoder model and how it is trained and evaluated. This section also describes the data used to test and train the architecture. In Sect. 4, we present and discuss the experimental results obtained from the model proposed in this study. We provide a detailed analysis of the dice coefficient, precision, and recall of the model. Finally, in Sect. 5, we conclude the study by summarizing the key findings and contributions of the proposed model. We also discuss potential avenues for future research and highlight the potential impact of the proposed model as a tool for skin lesion diagnosis.

2 Related Literature

For the early detection of skin cancer lesions using deep neural networks, Khan et al. [6] proposed training an encoder-decoder model on some image datasets. To prepare the datasets, the researchers implemented decorrelation formulation techniques to increase the visual intensity of the lesion region. A 3D box filter was also utilized to remove the noise produced when the decorrelation function was implemented. To segment the images from the background, the Mask R-CNN with the ResNet-50 used as the backbone architecture for the FPN was utilized. Otsu thresholding was then used to convert the RGB images to binary images. A DCNN which is a Dense-Net model was served as a feature extractor for the binary images. An entropy-controlled LS-SVM was also proposed as the classification technique for the extracted features.

To develop a robust DL algorithm for the detection of skin lesions, Bagheri, Tarokh, & Ziaratban [7] utilized the graph-based models to fuse the Retina-Deeplab and Mask R-CNN model together. The study also implemented the geodesic technique for combining the models. The models were then evaluated on three datasets which are the PH2, ISBI 2017, and DermQuest datasets. The proposed model was evaluated using the Jaccard Index metric and performed well with a performance of 80.4%. At the end of the study, two observations were made which are, firstly, the graph-based methods and geodesic methods outperformed other combination strategies utilized in the research. Secondly, the Retina-Deeplab segmentation approach has about 1% higher performance than the Mask-RCNN model when trained individually.

Hasan et al. [8] presented a DenseNet encoder-decoder model to boost the feature extraction capability and also decrease the problem of vanishing gradient in training the architecture. Two publicly available datasets were used for the performance evaluation of the architecture and the results obtained were found to be a mean IoU score of 77.5% on the ISIC-2017 datasets and 87.0% on the PH2 dataset respectively. This study also implemented a novel loss function for the optimization of the models.

The power of neural networks is their ability to perform well with less or no pre-processing techniques like image enhancements. A robust model should have the ability to correctly classify or segment different kinds of images with different resolutions. To develop a neural network robust at detecting images with varied resolutions, the study of [9] proposed a FrCNN which is a novel deep learning segmentation technique that preserves the features' resolution at each input pixel by removing all subsampling layers from the architecture. The model also used a transfer learning approach by using the pre-trained weights of the VGG-16 network layers and the images was also augmented to overcome the issue of insufficient training data. The objective function was further optimized with cross-entropy error function. The resultant function was then evaluated on the PH2 dataset and the ISBI 2017 dataset with a Jaccard Index of 84.9% and 77.11%. The researchers also concluded that their model outperforms state-of-art architectures like the U-Net and the SegNet model.

Research has shown that melanoma-type images usually have fuzzy boundaries and inhomogeneous features. To build a model robust enough effectively detect melanoma-type images. Another segmentation model was proposed by Lei, et al. [23] based on the FCNN to train a large network robust to the detection of melanoma skin cancer lesions. The study leveraged the residual skip connections of the ResNet architecture to a very large network based on the FCNN model. A PSI approach was proposed for the segmentation refinement. The model was evaluated on three sets of data which are mainly the 2016–2017 version of ISBI, and the PH2 dataset, and their results showed model recorded a dice coefficient of 85.66% on the ISBI 2017, 91.77% on the ISBI 2016, and 92.10% on the PH2 dataset. The researchers concluded that their model was more effective than other current methods for lesion segmentation. To enhance the pixel-level discriminative capability of the FCNN, Peng et al. [11] proposed a robust architecture by using stochastic weight averaging to prevent overfitting and also improve the generalization of the model. The suggested model was built on the U-Net with separable convolutional block. This gives the model the ability to capture semantic and context feature to improve its discriminative representation capabilities at the pixel-level. Evaluating the study on datasets from ISIC shows a significantly reduced complexity in addition to resolving the segmentation task.

Bektas et al. [12] also proposed a non-invasive framework which is a novel SRCRN based on Residual Network (ResNet) 50. In the first stage, high dimensional features are extracted which aid in achieving an accuracy of 95% on lesion segmentation. The ResNet-50 was used for the detection of melanoma in the second phase which also achieved 93% accuracy and an AUC of 97.3% at a 32 trials of batch size. This resulted in a 9.9% and 6.7% improvement in AUC and accuracy relative to other studies from the ISBI 2017 competition. The proposed framework used batch mode experiments and a class balancing strategy to enhance the training operation.

Tang et al. [13] introduced a two-stage multi-modal learning algorithm called FusionM4Net for multi-label skin lesion classification. The first stage employs a Fusion-Net that integrates clinical and dermoscopy image features and uses Fusion Scheme 1 for decision-level information fusion. In the second stage, patient metadata is incorporated using Fusion Scheme 2 to train an SVM cluster. Evaluating the FusionM4Net on the seven-point checklist dataset reveals the following: 74.9% for diagnostic tasks and 75.7% for multi-classification tasks without using the patient metadata while the entire FusionM4Net boosted accuracy to 77.0% and 78.5% respectively.

3 Methods and Materials

3.1 U-Net Design

This is a CNN encoder-decoder model, with the CNN serving as the encoder component to extract features from the source picture. To capture fine details of the image, the encoder component down samples the image and lowers the feature map resolution (Fig. 1).

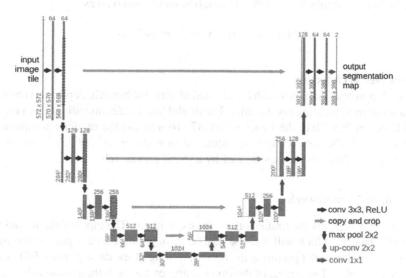

Fig. 1. Original structure of the U-Net architecture

Optimizers and Activation Functions

The Adam optimizer can be used in updating the weights in a neural network as denoted below:

$$w_t = w_{t-1} - \eta \frac{\hat{m}_t}{\sqrt{\hat{v}_t} + \varepsilon} \qquad (1)$$

The Rectified Linear Unit (ReLU) activation function is denoted as:

$$f(x) = max(0, x) \qquad (2)$$

$$f(x) = \begin{cases} 0 \; if \; x < 0 \\ x \; if \; x \geq 0 \end{cases} \qquad (3)$$

The Leaky ReLU activation function can be denoted as:

$$f(x) = \begin{cases} x \; if \; x > 0 \\ 0.01 \; x \; otherwise \end{cases} \qquad (4)$$

The Parametric ReLU activation function can be denoted as:

$$f(y_i) = y_i \; \text{if} \; y_i > 0 \tag{5}$$

$$f(y_i) = a_i y_i \; \text{if} \; y_i \leq 0 \tag{6}$$

where y_i is the input, and a_i is the negative slope which is the learnable parameter? This also means the PReLU becomes ReLU when $a_i = 0$ and Leaky ReLU when $a_i > 0$. Henceforth, the formula for the PReLU function can be denoted as:

$$f(y_i) = max(0, y_i) + a_i min(0, y_i) \tag{7}$$

Efficient-Net Architecture

Efficient-Net in brief is a CNN with evenly scaled network breadth, depth, and resolution using a set of preset scaling coefficients. The model has a different variant from the Base model Efficient-Net B0 to the EfficientNet B7. To increase the model's computational resources by 2^N, the network can be increased in width by α^N, β^N, γ^N, where $\alpha \geq 1$, $\beta \geq 1$, $\gamma \geq 1$ are constant determined by a small grid search.

3.2 Proposed Framework

Figure 2 below presents the framework of the experiment. The input of the model is the ISIC 2018 dataset. Which will then be passed through the encoder part of the model. The encoder of the model performs the downsampling of the dataset using EfficientNet pre-trained weights. The output of the down-sampled image is then passed through the decoder of the model to upsample the image and then get the segmented mask of the image using a 1×1 convolution at the last layer of the network.

The model consists of a pre-trained encoder network which is solely responsible for extracting features from a given input data. The decoder network on the other hand is used to generate the map for segmentation. Finally we have a skip connections that enable the fusion of multi-scale information. The pre-trained encoder model is fine-tuned using skin lesion images to adapt it to the segmentation task.

3.3 Dataset

Since neural networks are known to work with datasets that have undergone less or no preprocessing, the proposed model is assumed to perform well with no preprocessing methods like image enhancement, contrast enhancement, etc. To increase the speed of the model while training the size of the image was reduced to 255×255 pixels and also to ensure uniformity in the size of the image. The original images in the ISIC 2017 dataset had a shape of $(767 \times 1022 \times 3)$ pixels. The dataset also had ground truth images for both the model training and testing. The ground truth images unlike the training images are binary images that represent the mask of the original images.

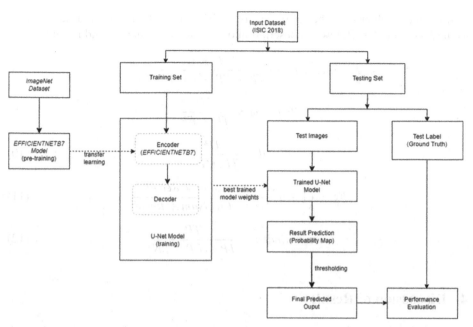

Fig. 2. Flow diagram of the proposed robust encoder-decoder model for effective segmentation of skin lesions.

3.4 Experimental Setup

Since transfer learning methods are known to perform very well on a small dataset, the study deduced that the size of the images for the training of the model was enough to train a good DL model therefore techniques like data augmentation used in maximizing the size of an image were not implemented on this dataset. To train the segmentation model on the dataset, the dataset was split into 3 samples, the data was first split into 80% for training and 20% for the testing of the model. The 80% of training was further split into 80% for the training and 20% for the model validation. In all, the number of testing, training, and validation images were found to be 518 images, 1558 images, and 518 images respectively. The training experiment also implemented a batch size of 8 and a learning rate of $1e^{-4}$ which was constant throughout the training and validation of all experiments. Some other hyper-parameters which were used interchangeably with others for the training process include the loss functions (DLs, FL, TL, FTL), activation functions like (ReLU, LeakyReLU, PReLU), and the Adam optimizer. These experiments will help us come up with the best pair that will be able to produce the best model for the segmentation.

3.5 Performance Metrics

Segmentation problems which form a part of classification problems and also the core problem for this study also have some performance metrics which are used to evaluate models build to solve certain segmentation problems.Among these we have: the Dice

Coefficient which can also be referred to as the F1 score., The IoU score is also known as the Jaccard Index. Below is a description of certain performances and formulae.

$$Accuracy = \frac{TP + TN}{TP + FP + TN + FN} \tag{8}$$

$$Precision = \frac{TP}{TP + FP} \tag{9}$$

$$Recall = \frac{TP}{TP + FN} \tag{10}$$

$$F_\alpha = \left(1 + \alpha^2\right) \frac{Precision \times Recall}{\alpha^2 \times Precision + Recall} \tag{11}$$

$$JaccardIndex = \frac{TP}{TP + FP + FN} \tag{12}$$

4 Discussion of Results

The ReLU activation function which has been the de facto standard to reduce the vanishing gradient problem in a neural network was also utilized in the model to ascertain its performance. Using the ReLU activation and the Adam optimizer to optimize the various loss functions. Using the Dice Loss function to train the model, it was observed that the model obtained dice score, Jaccard Index/IoU, Precision, and Recall of 88%, 81%, 90.7%, and 89.2% respectively. It was also observed that training the model with a Tversky Loss function, the model had a dice score of 90.1%, an IoU of 83.5%, and a precision and recall of 91.7% and 91.5% respectively. The experiment performed using the Focal Tversky Loss function was observed to have a score of 90.8%, 90.1%, 83.47%, and 91.7% for precision, DLs, IoU, and recall respectively. Optimizing the model with the focal Loss function converged faster with an epoch of 13 but obtained the lowest performance dice coefficient score of 87.6%, IoU score of 79.9%, a precision of 90.98%, and recall of 87.03%.

The Leaky ReLU activation function was also introduced into the model and used the Adam optimizer to optimize the 4 loss functions which are the Tversky, Focal Tversky, Focal, and Dice loss functions. It was observed after the experiment that, the trained model on the DLs function achieved a precision of 90.3%, a dice coefficient score of 90.2%, a precision of 90.3%, an IoU score of 83.6%, and a recall of 92.5%. The model was also observed to converge on the 36th epoch. The model trained to optimize the Tversky Loss function also converged on the 27th epoch and obtained dice score, IoU, precision, and recall of 89.9%, 83.2%, 89.5%, and 92.5% respectively. Using the Focal Tversky Loss function as the cost for the model with the Adam optimizer and Leaky ReLU activation function, it was observed that, the model's performance for dice coefficient, IoU, precision, and recall was recorded to be 90.1%, 83.5%,92.1%, 90.7% respectively. Optimizing the Focal Loss with the Adam optimizer was the experiment that obtained the least performance but converged faster with an epoch of 16. The model obtained a dice coefficient of 88.7%, an IoU of 81.1%, and a precision and recall of 89.9% and 90.4% respectively.

It can also be observed from the study that, the experiment which achieved the highest performance was using the PReLU activation function in the model and the Adam optimizer to optimize the TL function. The dice score, IoU score, precision, and recall for this experiment were recorded to be 90.76%, 83.67%, 91.1%, and 91.85% respectively. The model also converged on the 36th epoch during training. The PReLU activation function was also utilized to optimize the Focal Tversky Loss and recorded a precision of 90.5%, a DC of 90%, an IoU of 83.6%, and a recall of 91.9%. The model also converged on the 50th epoch. The results obtained when the PReLU activation was used in the model and the Adam optimizer was used in optimizing the FL function metric scores of 88.5%, 81.1%, 88.9%, 91.4% for DC, IoU, precision, and recall respectively, and converged on the 16th epoch. It can be observed that optimizing the focal loss function causes the model to converge faster and thereby a lesser performance which can be attributed to the model getting stuck in a local minimum during the training to optimize the loss function. The PReLU activation function was also utilized in the model and using the Adam optimizer to optimize the Dice Loss function, the model obtained a performance of 90.8%, 90.1%, 83.6%, 91.7% for precision, DLs, IOU, and recall respectively (Table 1 and Fig. 3).

Table 1. Metrics for ReLU, LeakyReLu, and PreLu activation functions

		Dice	IOU	Precision	Recall
ReLu	Dice Loss	0.88338	0.81334	0.90746	0.89182
	Focal Tversky Loss	0.90093	0.83467	0.90829	0.91733
	Tversky Loss	0.90108	0.83509	0.91079	0.91543
	Focal Loss	0.87643	0.79916	0.90977	0.87028
LeakyReLU	Dice Loss	0.90260	0.83596	0.90320	0.92521
	Focal Tversky Loss	0.90144	0.83512	0.92083	0.90663
	Tversky Loss	0.89945	0.83203	0.89535	0.92881
	Focal Loss	0.88479	0.81102	0.89938	0.90389
PReLU	Dice Loss	0.90136	0.83555	0.90813	0.917172
	Focal Tversky Loss	0.90035	0.83345	0.904980	0.91960
	Tversky Loss	**0.90767**	**0.83672**	**0.91131**	**0.91858**
	Focal Loss	0.88491	0.81109	0.88893	0.91482

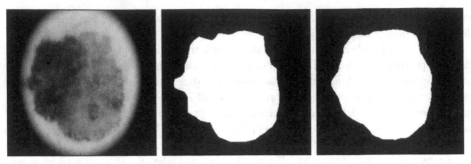

Fig. 3. Original Data (left), Expected Segmentation (middle), Model's Segmented Image (Right)

5 Conclusion

In conclusion, this research paper aimed to optimize the performance of a transfer learning based U-Net architecture used for segmenting skin lesions. In the study, the Efficient-Net architecture with ReLU activation function and dice loss was used as the baseline model for the experiment. This research study is divided into two parts, the first part of the study optimized the performance of the U-Net architecture by replacing the encoder of the architecture with the pre-trained Efficient-Net model. The pre-trained model's weights were used as the initial weights for the parameters of the encoder. Batch normalization was introduced into the architecture to normalize the input of each layer, effectively reducing internal covariant shift. The baseline architecture achieved a performance with a DC of 90% which outperforms the classical U-Net architecture (86%). The second part of the study was to experiment and identify the best combination of hyperparameters to optimize to model for the efficient segmentation of skin lesions. The hyper-parameter which was considered for the experiment was the activation function and the loss function with the results captured in the analysis of this paper. It was remarked that initializing the encoder of the U-Net function with the pre-trained weights of the Efficient-Net B7 model and optimizing the model with Tversky loss function with PReLU activation function produces a better performance with a dice score of 90.8%, recall of 91.1% and precision of 91.8% and PReLU activation function to optimize the Tversky loss function produced achieved a high dice score. In future works, similar techniques can be applied to variants of the U-Net architecture like the Nested-UNet, and Double-UNet to check their segmentation's performance on skin lesions.

Abreviations

CNNs: Convolutional Neural Networks	FTL: Focal Tversky Loss
DCNN: Deep Convolutional Neural Network	IoU: Intersection over Union
DC: Dice Coefficient	LS-SVM: Least Square Support Vector Machine
DL: Deep Learning	PSI: Probability Step-wise Integration

(continued)

(*continued*)

CNNs: Convolutional Neural Networks	FTL: Focal Tversky Loss
DLs: Dice Loss	R-CNN: Mask-Region-Based Convolutional Neural Network
FL: Focal Loss	ResNet: Residual Neural Network
FPN: Feature Pyramid Network	TL: Tversky Loss

References

1. Arasu, A., Meah, N., Sinclair, R.: Skin checks in primary care. Aust. J. Gener. Pract. **48**(9), 614–619 (2019)
2. Kim, C.-I., Hwang, S.-M., Park, E.-B., Won, C.-H., Lee, J.-H.: Computer-aided diagnosis algorithm for classification of malignant melanoma using deep neural networks. Sensors **21**(16), 5551 (2021)
3. Jayatilake, S.M.D.A.C., Ganegoda, G.U.: Involvement of machine learning tools in healthcare decision making. J. Healthc. Eng. (2021)
4. Yamashita, R., Nishio, M., Do, R.K.G., Togashi, K.: Convolutional neural networks: an overview and application in radiology. Insights Imaging **9**, 611–629 (2018)
5. Khan, M.A., Akram, T., Zhang, Y.-D., Sharif, M.: Attributes based skin lesion detection and recognition: a mask RCNN and transfer learning-based deep learning framework. Pattern Recognit. Lett. **143**, 58–66 (2021)
6. Bagheri, F., Tarokh, M.J., Ziaratban, M.: Skin lesion segmentation from dermoscopic images by using Mask R-CNN, Retina-Deeplab, and graph-based methods. Biomed. Signal Process. Control **67**, 102533 (2021)
7. Hasan, K., Dahal, L., Samarakoon, P.N., Tushar, F.I., Marti, R.: DSNet: automatic dermoscopic skin lesion segmentation. Comput. Biol. Med. **120**, 103738 (2020)
8. Al-masni, M.A., Al-antari, M.A., Choi, M.-T., Han, S.-M., Kim, T.-S.: Skin lesion segmentation in dermoscopy images via deep full resolution convolutional networks. Comput. Methods Programs Biomed. **162**, 221–231 (2018). https://doi.org/10.1016/j.cmpb.2018.05.027
9. Peng, T., et al.: Efficient skin lesion segmentation using separable-Unet with stochastic weight averaging. Comput. Methods Programs Biomed. **178**, 289–301 (2019)
10. Bektas, J., Bektas, Y., Kangal, E.E.: Integrating a novel SRCRN network for segmentation with representative batch-mode experiments for detecting melanoma. Biomed. Signal Process. Control **71**, 103218 (2022)
11. Tang, P., Yan, X., Nan, Y., Xiang, S., Krammer, S., Lasser, T.: FusionM4Net: a multi-stage multi-modal learning algorithm for multi-label skin lesion classification. Med. Image Anal. **76**, 102307 (2022)
12. Narayanamurthy, V., et al.: Skin cancer detection using non-invasive techniques. RSC Adv. **8**(49), 28095–28130 (2018)
13. Apalla, Z., Lallas, A., Sotiriou, E., Lazaridou, E., Ioannides, D.: Epidemiological trends in skin cancer. Dermatol. Pract. Concept **7**(2), 1 (2017). https://doi.org/10.5826/dpc.0702a01
14. Shen, D., Wu, G., Suk, H.-I.: Deep learning in medical image analysis. Annu. Rev. Biomed. Eng. **19**, 221–248 (2017)
15. Lei, B., et al.: Skin lesion segmentation via generative adversarial networks with dual discriminators. Med. Image Anal. **64**, 101716 (2020)

16. Wu, H., Chen, S., Chen, G., Wang, W., Lei, B., Wen, Z.: FAT-Net: feature adaptive transformers for automated skin lesion segmentation. Med. Image Anal. **76**, 102327 (2022)
17. Baghersalimi, S., Bozorgtabar, B., Schmid-Saugeon, P., Ekenel, H.K., Thiran, J.-P.: DermoNet: densely linked convolutional neural network for efficient skin lesion segmentation. EURASIP J. Image Video Process. **1**, 1–10 (2019)

A System for Liver Tumor Detection

Anuradha Thakare[1]([⊠]) [iD], Shreya Pillai[1] [iD], Rutuja Nemane[1] [iD], Nupur Shiturkar[1] [iD], Anjitha Nair[1] [iD], and Ahmed M. Anter[2] [iD]

[1] Pimpri Chinchwad College of Engineering, Pune, India
anuradha.thakare@pccoepune.org, shreyapillai29@gmail.com,
rutuja.b.nemane2808@gmail.com, shiturkarnupur@gmail.com,
anjithanair.146@gmail.com
[2] Faculty of Computers and Artificial Intelligence, Beni-Suef University, Beni Suef, Egypt
sw_anter@yahoo.com

Abstract. Liver cancer is found to be sixth prevalent cancer globally and is one of the vital reasons of death. The radiologist has to confirm the diagnosis and liver segmentation manually, which is a highly complex process. Thus, it is important to diagnose liver tumors in the primitive stage, but it is a complex process due to some problems of CT scans like noise, annotation, low resolution, and closely connected organs making it difficult to segment out different organs. Here, we put forth an automated method for the liver segmentation and tumor detection. Considering the fact that there is a shortage of training data in medical imaging, we have opted for the transfer learning (TL) technique. It is a popular technique when it comes to medical imaging datasets since it provides efficient results even with small datasets. In preprocessing, gaussian filter is used to eliminate noise from the images. In liver segmentation phase from the abdominal CT scans, Unet-VGG16 model is used which has minimum value of loss and maximum value of accuracy. Here, we have compared our results using VGG U-Net and SegNet models for liver segmentation from abdominal CT scan which achieved a true positive rate of 73% and 92%, respectively. Moreover, the proposed approach using transfer learning achieved validation accuracy of 83% for liver tumor detection.

Keywords: Liver tumor · VGG U-Net · SegNet · Deep convolutional neural network · Transfer learning · Inception V3 · Segmentation · Classification · Augmentation · CT images · Computer aided diagnosis

1 Introduction

Liver is most crucial, and largest solid organ in human body. It performs some major functions like removing toxins from the blood, maintaining proper blood sugar levels, regulating blood clotting, and numerous other vital functions. The liver is vulnerable to many diseases, which if not diagnosed in the primitive stage, may lead to some serious consequences. The liver disease may be genetic or even due to the lifestyle carried by one. The liver is vulnerable to many diseases which may lead to liver tumor. It is very crucial for these tumors to be detected and diagnosed at an early stage and the previously used methods take too long for detection and are not accurate with many false positive regions. Automated techniques for detection will not only be time-efficient but also aid the surgeons to take action at an early stage itself. Hence, there is a need for

J. Abawajy et al. (Eds.): ICICCT 2023, CCIS 1841, pp. 129–141, 2023.
https://doi.org/10.1007/978-3-031-43838-7_10

modern automated methods for the diagnosis of these tumors which will help to speed up the treatment and also increase efficiency. In the field of medical imaging, computed tomography (CT) images are popular tool for detection and diagnosis purposes [4].

In medical image segmentation, transfer learning is better than other classical methods because there is a scarcity of data. For segmentation, various methods like U-Net and SegNet are used which have pre-trained CNN models. We have selected these methods because they are used in various medical applications [6] and they could achieve a good accuracy.

This paper discusses an automated solution for tumor detection from CT scans. TL technique has been known to provide accurate results even with less dataset whereas the other techniques usually fail to do so. We have focused on different methods for liver segmentation in this paper. Our proposed approach has better results in case of small data.

In this paper, Sect. 2 is a survey of the previous works from this domain. Section 3 shows our proposed model that can be used for liver tumor detection. The experimental results are shown in Sect. 4. We have summarized our conclusion and the aim of future work in Sect. 5.

2 Literature Review

Modified U-Net and a novel class balancing algorithm were suggested in a paper for segmentation of liver tumor through CT images. The algorithm proposed here is obtained from the standard U-Net architecture, along with some modifications for enhancing the segmentation performance. Firstly, the number of filters were reduced for all the convolutional blocks, along with a new batch normalization and a dropout layer post every convolutional block. Due to all these modifications in the standard U-Net architecture, the model showed better results. But, the model needs certain improvements in the segmentation of small and irregular tumors [2].

A work was published wherein a stack of CT scans of the chest are analyzed. Novel approaches for segmenting multimodal grayscale lung CT scans are presented. The desired regions of interest (ROI) are discovered in traditional approaches utilizing the MGRF model. The outcomes of the suggested FCM and Convolutional Neural Network based model are compared to the outcomes of the traditional Markov-Gibbs Random Field model based method. The final results showed that the suggested method is capable of precisely segmenting various types of complicated multimodal medical images [4].

Another work was published wherein a fully automatic technique is offered for successfully segmenting lungs from thoracic CT scans. Here, the segmentation of the lungs are divided into two stages: DCNN model-based lung segmentation and a two-pass contour refining of local and global. Here, the contours of the lung were rectified and the algorithm achieved higher accuracy [5].

In order to segregate liver tumors from CT scans, a paper from 2018 recommends combining neutrosophic sets (NS) with fast fuzzy c-means and an adaptive watershed technique. The reason for integrating three methods is that one method overcomes the drawbacks of the other method, which improves the overall results. The proposed method consisted of five phases: Pre-processing, CT image transformation to NS domain, Post-processing, Liver parenchyma segmentation, and Tumors segmentation and extraction.

The applied technique shows that indeterminacy and uncertainty are handled more efficiently, lowers over-segmentation and gives good results even with noisy data. The accuracy achieved by the model was almost 95% of good liver segmentation [13].

In 2022, an article was published which discussed the use of the ResU-Net architecture. They have also used residual blocks in this model which help in collecting large amount of data. They have used Hounsfield Windowing unit for measuring the density of tissue and getting proper visuals of the organ. Histogram equalization was used for normalization of the image. The segmentation of the tumor and liver using the ResU-Net model produced the best results out of all the alternatives [18].

In 2022, a paper was published in which the authors have tried to build a system that segments liver and tumor from CT images [19]. It is an automated system which uses Deep Learning Algorithm. In this paper, modified ResUNet is used to for the segmentation of the liver and the tumours. They have used the LITS dataset which contains three-dimensional images. These images contain no distinct mask for the tumours. For segmenting the liver tumours, CNN model is used. Further, they have also mentioned that more dataset and variety of pre-processing techniques can bring in more accurate results.

Table 1 shows different studies on liver tumor segmentation with accuracies and results achieved.

Table 1. Comparison of other related work.

Authors	Results				Findings
	DSC	Precision	Recall	CCR	
Wen Li et al.	80.06% ± 1.63%	82.67% ± 1.43%	84.34% ± 1.61%	–	CNN's perform better segmentation than the traditional machine learning algorithms
Pravitasari, Anindya et al.	–	–	–	95.69%	UNet-VGG16 has greater performance compared to the U-Net model for segmentation
Ayalew, Y.A. et al.	Segmentation: 0.96 Tumor Detection: 0.74	–	–	–	Segmentation improved by reducing class imbalance

(*continued*)

Table 1. (*continued*)

Authors	Results				Findings
	DSC	Precision	Recall	CCR	
Jalal Deen K. et al.	Using MGRF Model: 97% Using FCM+CNN: 99%	–	–	–	Used FCM clustring for computing every empirical dispersion of the image. This gave proper boundaries of the regions
Caixia Liu et al.	97.95%	–	–	–	Edge direction tracing technique improved accuracy of the segmentation model

3 Proposed Model

Figure 1 depicts our proposed model for liver tumor detection. The steps involved are: pre-processing, augmentation, liver segmentation, and finally the liver tumor detection and classification as malignant or benign tumor. The abdominal CT image is passed to the automated system and then the above steps are performed. More details about these steps are as follows:

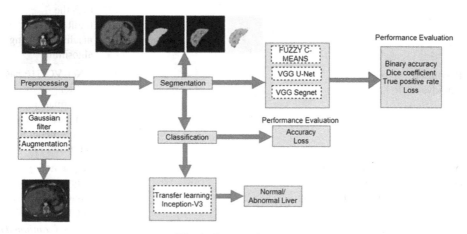

Fig. 1. Proposed Model

3.1 Preprocessing

Before segmenting the liver and liver tumor parts, the removal of noise from CT images is crucial. So, one of the linear filters called a Gaussian filter is used. It is used for smoothening the images and removing detail and noise from the images. It is more effective in blurring the images. When we are working with images, we require a two-dimensional gaussian function. It is the product of two one-dimensional Gaussian functions which is given by:

$$G(x, y) = \frac{1}{2\pi\sigma^2} e^{-\frac{x^2+y^2}{2\sigma^2}}$$
(1)

Here, σ represents the standard deviation. Gaussian filter is faster than median filter as multiplication and addition is faster than sorting.

3.2 Augmentation

Augmentation is very useful for training models. It is used to artificially increase the dataset by slightly modifying the dataset or by adding more existing data. The model overfits if the dataset is small and augmentation reduces overfitting.

The dataset that we used had less number of images, so we performed augmentation. We rotated the images by 5° and also flipped the images horizontally to increase the abdominal CT images. The images were not flipped vertically as the liver images can never be upside-down and hence flipping it vertically would generate wrong images.

3.3 Segmentation

Segmentation is done so that the regions of interest can be captured effectively. First, Fuzzy C-means (FCM) algorithm is used for forming clusters out of the image. The FCM is used to divide the image into two or more clusters which will help for further segmentation tasks.

Then the clustered images from FCM algorithm are passed to the segmentation process. In this paper, we are discussing two of the methods that can be used for segmentation.

VGG16 U-Net: For computer aided diagnosis, segmentation of the parts needs to be accurate. CNN is widely used in segmentation but it may lead to overfitting. So CNN is further expanded into an U-Net architecture mainly for segmentation of biomedical images. This can be used for not only classification but also knowing the affected areas in the images.

U-Net can also be used along with pre-trained encoder weights. One of them is VGG-16 encoder. The architecture of U-Net can be changed by replacing encoder part with VGG 16 network. The VGG 16 network is pre-trained on top of ImageNet dataset. It helps in reducing overfitting.

Segnet Model: Another semantic segmentation model is SegNet. It consists of an encoder. This is constructed with 13 layers identical to the VGG16 network. This model is small and can be trained easily as compared to other models because full connected layers from VGG-16 are removed in this architecture. So, this architecture is convolutional only.

Here, a decoder uses the indices of transferred pool from its encoder for up sampling its input, that will further produce sparse feature maps. For densifying the feature map, convolution with a trainable filter bank is performed. For pixel-by-pixel classification, the soft-max classifier is employed.

3.4 Classification Model

The segmented image is further classified into different categories to get the required output. For the classification of images, we can use different intelligent learning techniques. Convolutional neural networks result in overfitting when the amount of data is small and are computationally expensive. Transfer learning proves to be overcoming these problems. In transfer learning, pre-trained CNN model is used. We have used Inception V3 model which is trained using ImageNet dataset. Here, we just use the top layers for our classification part and the other below layers are frozen. This helps us to focus on the classification part instead of the entire model creation and training. This implies that transfer learning shows great improvements in the speed of classification. As these models work with fewer data too, overfitting is avoided which results in improved accuracy. For further reduction of overfitting, dropout is also used.

3.5 Performance Measures

To prove the results achieved we used different common performance measures such as:

Loss: It measures the performance of the model and gives a value between 0 and 1. The performance of the model is indirectly proportional to the loss.

Dice Coef: This performance measure is mainly for segmentation problems. It measures similarity between two sets of data by matching the value as well as the position.

$$\text{Dice Score} = \frac{2 \cdot |A \cap B|}{|A| + |B|} \tag{2}$$

Binary Accuracy: It calculates the percentage of predicted values that matches the actual values.

$$\text{Binary accuracy} = \frac{\text{Number of accurately predicted records}}{\text{Total number of records}} \tag{3}$$

True Positive Rate: It's rate of actual positives predicted correctly.

$$\text{True positive rate} = \frac{\text{True positives}}{\text{True positives} + \text{False Negatives}} \tag{4}$$

4   Results with Discussion

4.1   Dataset Description

Our dataset is composed of 75 abdominal CT scans for male and female with a diagnosis report for each patient. This dataset contains different variations of complexity. The resolution of the images is 630 × 630 with 72 DPI. Along with the CT images, their masks are also included in the dataset which are used in the segmentation process. Figure 2 displays samples of CT images of abdominal part. As per given figure, segmentation task is not easy as the intensity of all the abdominal organs in the image is same and the liver is highly connected to other tissues.

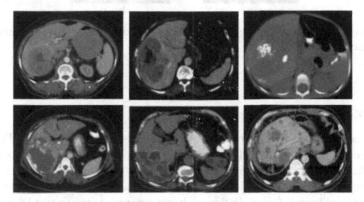

Fig. 2. Dataset Samples.

4.2   Computer Settings and Parameters

All the implementation is done on google Colab using Python3 and Tensorflow library. The PC used for implementing the modules has an intel core I5-8265U processor with 1.6 Ghz base frequency. The FCM separates the liver region from CT scan of the abdomen image with 3 clusters. The clustered image is then passed to the VGG 16 seg-net model for segmentation [2].

4.3 Preprocessing

The CT images shown in Fig. 2 are passed to a gaussian filter for removing noise and blurring the images. Figure 3 depicts the input original image to the Gaussian filter and its output on applying the filter.

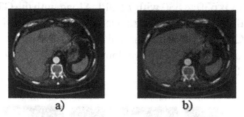

a)   b)

Fig. 3. Results of the preprocessing phase. (a) Original abdominal CT scan and (b) Output of gaussian filter.

4.4 Liver Segmentation

For liver segmentation, first the pre-processed image is passed through the FCM algorithm to cluster the abdominal CT images into parts. Following Fig. 4 is the output after applying fuzzy c-means clustering.

The clustered image is provided as an input to the deep learning (DL) model for segmenting liver from CT scan. In this study we compare between two DL techniques with transfer learning tuning VGG16 U-Net and SegNet models.

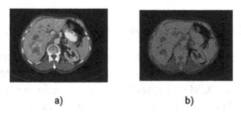

a)   b)

Fig. 4. Results of the preprocessing phase. (a) Original abdominal CT scan and (b) Output of gaussian filter.

VGG16 U-Net: The clustered image is fed to the U-net model for segmentation. Figure 5 represents the output of segmentation using VGG 16 U-Net model.

SegNet Model: Following Fig. 6 shows the output when clustered images are fed to the Segnet model for liver segmentation.

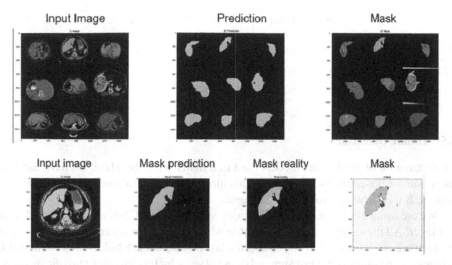

Fig. 5. Segmentation output of VGG16 U-Net model

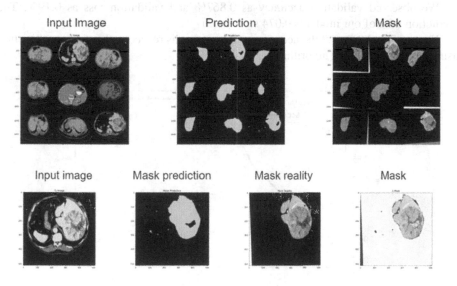

Fig. 6. Segmentation output of SegNet model

The overall performance measures for the methods used are shown below (Table 2):

Table 2. Comparison between u-net and segnet

Methods	Loss	Dice Coef	Binary Accuracy	True positive rate
U-Net Model	−0.60	0.60	0.85	0.73
SegNet Model	−0.64	0.64	0.87	0.92

4.5 Tumor Detection

There are a lot of pre-trained models used in transfer learning. Here, we have used the Inception V3 pre-trained CNN model. This model has been trained using the ImageNet dataset. It uses features from the CT images.

We are using all layers from the Inception v3 pre-trained CNN model, except for the last layer. All these layers are non-trainable which reduce the overfitting. On the top of the network, we have used 2 fully connected layers, one has 128 hidden units with ReLU activation, and other is the last sigmoid dense layer which is used for classification. We used an RMSprop optimizer having a learning rate of 0.0001.

We observed validation accuracy as 0.867% and validation loss as 0.19%. The prediction time of our model is 0.074 s.

Figures 7 and 8 depict the accuracy and loss graphs respectively of tumor detection using Transfer Learning algorithm.

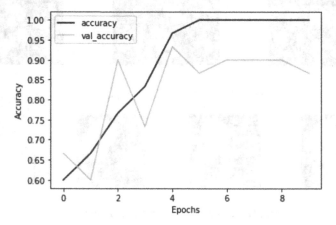

Fig. 7. Training vs. Validation accuracy for classification

Figure 9 shows the classification matrix of our model.

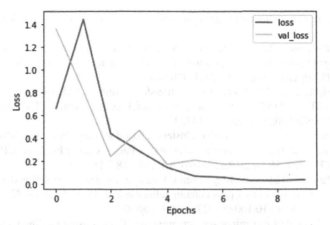

Fig. 8. Training vs. Validation loss for classification

	precision	recall	f1-score
Normal	0.84	0.86	0.76
Tumor	0.86	0.84	0.76
accuracy			0.76
macro avg	0.80	0.70	0.76
weighted avg	0.80	0.76	0.76

Fig. 9. Classification matrix

5 Conclusion and Future Work

The abdominal CT images are pre-processed for the removal of noise from CT images and blurring the images using gaussian filter. First, we formed the clusters of images using Fuzzy C-means. This helps in further process of segmentation. Then we used two methods for segmentation (VGG U-Net and VGG Segnet) and compared them. The Segnet model shows better accuracy than the U-Net model. Moreover, the segnet model is simpler, smaller and needs less computation power. For classification, as the amount of data is less, CNN models are overfitting. Transfer learning avoids this overfitting and gives better accuracy.

Further, our aim is to enhance the accuracy of the model by collecting more datasets and using GANs for synthetic medical image augmentation for improving the performance.

References

1. Li, W., Jia, F., Hu, Q.: Automatic segmentation of liver tumor in CT images with deep convolutional neural networks. J. Comput. Commun. (2015).

https://www.scirp.org/journal/paperinformation.aspx?paperid=61314 . Accessed 17 Mar 2022

2. Ayalew, Y.A., Fante, K.A., Mohammed, M.: Modified U-Net for liver cancer segmentation from computed tomography images with a new class balancing method. BMC Biomed. Eng. **3**, 4 (2021). https://doi.org/10.1186/s42490-021-00050-y

3. Pravitasari, A., et al.: UNet-VGG16 with transfer learning for MRI-based brain tumor segmentation. TELKOMNIKA Telecommun. Comput. Electron. Control **18**, 1310 (2020). https://doi.org/10.12928/telkomnika.v18i3.14753

4. Jalal Deen, K., et al.: Fuzzy-C-means clustering based segmentation and CNN-classification for accurate segmentation of lung nodules. Asian Pac. J. Cancer Prev. APJCP **18**(7), 1869–1874 (2017). https://doi.org/10.22034/APJCP.2017.18.7.1869

5. Liu, C., Pang, M.: Extracting lungs from CT images via deep convolutional neural network based segmentation and two-pass contour refinement. J. Digit. Imaging **33**(6), 1465–1478 (2020). https://doi.org/10.1007/s10278-020-00388-0

6. Almotairi, S., et al.: Liver tumor segmentation in CT scans using modified SegNet. Sensors **20**(5), 1516 (2020). https://doi.org/10.3390/s20051516

7. Iriawan, N., Pravitasari, A., Fithriasari, K., Irhamah, Purnami, S., Ferriastuti, W.: Comparative study of brain tumor segmentation using different segmentation techniques in handling noise, pp. 289–293 (2018). https://doi.org/10.1109/CENIM.2018.8711004

8. Hesamian, M.H., Jia, W., He, X., et al.: Deep learning techniques for medical image segmentation: achievements and challenges. J. Digit. Imaging **32**, 582–596 (2019)

9. Anter, A.M., Bhattacharyya, S., Zhang, Z.: Multi-stage fuzzy swarm intelligence for automatic hepatic lesion segmentation from CT scans. Appl. Soft Comput. **96**, 106677 (2020). ISSN 1568-4946. https://doi.org/10.1016/j.asoc.2020.106677

10. Salem, M.A.-M., Atef, A., Salah, A., Shams, M.: Recent survey on medical image segmentation. Comput. Vis. Concepts Methodol. Tools Appl., 129–169 (2018). https://doi.org/10.4018/978-1-5225-5204-8.ch006

11. Ronneberger, O., Fischer, P., Brox, T.: U-Net: convolutional networks for biomedical image segmentation. arXiv. abs/1505.04597 (2015)

12. Anter, A.M., Azar, A.T., Hassanien, A.E., El-Bendary, N., ElSoud, M.A.: Automatic computer aided segmentation for liver and hepatic lesions using hybrid segmentations techniques. In: 2013 Federated Conference on Computer Science and Information Systems, pp. 193–198 (2013)

13. Anter, A.M., Hassenian, A.E.: CT liver tumor segmentation hybrid approach using neutrosophic sets, fast fuzzy c-means and adaptive watershed algorithm. Artif. Intell. Med. **97**, 105–117 (2019). https://doi.org/10.1016/j.artmed.2018.11.007. Epub 2018 Dec 14. PMID: 30558825

14. Anter, A.M., Hassanien, A.E., ElSoud, M.A., Kim, T.:Feature selection approach based on social spider algorithm: case study on abdominal CT liver tumor. In: 2015 Seventh International Conference on Advanced Communication and Networking (ACN), pp. 89–94 (2015). https://doi.org/10.1109/ACN.2015.32

15. Nemane, R., Thakare, A., Pillai, S., Shiturkar, N., Nair, A.: Comparative analysis of intelligent learning techniques for diagnosis of liver tumor from CT images. In: Verma, P., Charan, C., Fernando, X., Ganesan, S. (eds.) Advances in Data Computing, Communication and Security. LNDECT, vol. 106, pp. 27–37. Springer, Singapore (2022). https://doi.org/10.1007/978-981-16-8403-6_3

16. Katre, P.R., Thakare, A.: Detection of lung cancer stages using image processing and data classification techniques. In: 2017 2nd International Conference for Convergence in Technology (I2CT), pp. 402–404 (2017). https://doi.org/10.1109/I2CT.2017.8226160

17. Anter, A.M., Abu ElSoud, M., Azar, A.T., Hassanien, A.E.: A hybrid approach to diagnosis of hepatic tumors in computed tomography images. Int. J. Rough Sets Data Anal. **1**(2), 31–48 (2014)
18. Sabir, M.W., et al.: Segmentation of liver tumor in CT scan using ResU-Net. Appl. Sci. **12**, 8650 (2022). https://doi.org/10.3390/app12178650
19. Manjunath, R.V., Kwadiki, K.: Automatic liver and tumour segmentation from CT images using deep learning algorithm. Results Control Optim. **6**, 100087 (2022). ISSN 2666-7207. https://doi.org/10.1016/j.rico.2021.100087

High Speed and Efficient Reconfigurable Histogram Equalisation Architecture for Image Contrast Enhancement

P. Agalya[1]($\boxtimes$) (iD) and M. C. Hanumantharaju[2] (iD)

[1] Department of Electronics and Communication Engineering, Sapthagiri College of Engineering, Bangalore, India
agalyas@sapthagiri.edu.in
[2] Department of Electronics and Communication Engineering, BMS Institute of Technology and Management, Bangalore, India
mchanumantharaju@bmsit.in

Abstract. Histogram equalisation is a point processing technique used to improve image quality in wide variety of applications including real-time video surveillance, medical imaging, industrial automation, intelligent self-navigation systems, and oceanography. This research paper proposes a unique, simple high-speed reconfigurable architecture for the implementation of histogram equalisation on field programmable gate arrays (FPGA). The proposed histogram equalisation architecture comprises of a pixel mapping unit and a histogram computing unit. The histogram computation unit makes use of counters and comparators to compute the cumulative histogram of images. The pixel mapping unit comprise of a delay control unit with a register array to map the output pixel intensity value for each input pixel grey level by adopting a straightforward and effective pixel transformation method. The proposed architecture has been developed using register transfer level (RTL) - compliant Verilog HDL code, simulated using Xilinx's integrated simulator (ISim), synthesized and implemented on a Kintex7 family of Field Programmable Gate Array (FPGA) device. Simulation and synthesis results demonstrate that the proposed architecture can be implemented with a processing time of 2.353 ns and a maximum frequency of operation of 425 MHz. Experimental results shows that the proposed method outperforms other existing methods.

Keywords: Histogram Equalisation · Field Programmable Gate Array · Reconfigurable Architecture · low contrast images

1 Introduction

Image enhancement is a preprocessing technique used to improve an image's quality to make it more aesthetically pleasing and to provide better input for image classification. Spatial and frequency domain techniques are the two broad categories of image enhancement methods. Histogram equalization is one of the point-processing spatial domain techniques used in the field of medicine for image registration using mutual

J. Abawajy et al. (Eds.): ICICCT 2023, CCIS 1841, pp. 142–156, 2023.
https://doi.org/10.1007/978-3-031-43838-7_11

information (MI) computation, contrast enhancement, image compression and particle filters for video object tracking [9]. The basic idea of histogram equalization is to give more weighted emphasis to the dominant pixel grayscale levels which occur more frequently and less weighted emphasis to the grayscale levels which occur less frequently in an image [5]. Histogram equalisation technique aims to improve the contrast of an image by spreading the histogram over a wide range of grayscale levels.

Histogramcequalization improves the contrast of an image in four steps as shown in Eqs. (1), (2) and (3):

(i) Compute the frequency of occurrence of each pixel grayscale level in the image (nk).
(ii) Compute the probability density function (PDF)
(iii) Compute the cumulative distribution function (CDF)
(iv) Apply the required intensity transformation function.

$$PDF(k) = for\ k = 0\ to\ 255 \tag{1}$$

$$CDF(r) = \sum_{k=0}^{r} PDF(k)\ for\ r = 0\ to\ 255 \tag{2}$$

$$T(i) = X_{L-1} \times CDF(i)\ for\ i = 0\ to\ 255 \tag{3}$$

The term n_k refers to the number of pixels with grayscale level value equal to k in an image, where the values of k and r range from 0 to 255. In addition PDF refers to probability density function, CDF refers to the cumulative distribution function, X_{L-1} refers to maximum 8 bit intensity value and T(.) is the transformation function.

Generally natural scenery and medical images will have more of background image pixels than the actual object pixels. Histogram equalisation over enhances both background and information pixels. As the histogram equalisation technique is simple to implement, various researchers have proposed different versions of histogram equalisation techniques [7] such as Brightness preserving Bi Histogram Equalization (BBHE) proposed by Kim et al. [1], Dualistic Sub Image Histogram Equalization (DSIHE) by Wang et al. [3], Minimum Mean Brightness Error BI Histogram Equalization (MMBEBHE) by Chen et al. [4] divide an image into two sub images based on mean, median and the mean brightness difference and apply histogram equalisation locally on each sub image.

As real time applications demand less computational time and utilization of fewer hardware resources, many researchers have found it more difficult to implement the aforementioned strategies in real-time image processing applications. Utilizing a Field Programmable Gate Array (FPGA) to implement histogram equalisation techniques will speed up the computation and save resources because FPGA implementation provides parallelism and pipelining while processing the image data.

This research paper is organized as follows: Sect. 2 describes the related work. Section 3 presents the proposed architecture and its implementation. Section 4 discusses the experimental results and analysis of proposed architecture design and Sect. 5 is presented with a brief conclusion.

2 Related Work

The huge challenge in hardware implementation of histogram equalization algorithm is to reduce the histogram statistic computation time by utilizing fewer hardware resources. The two most convenient architectures proposed in the literature for computing histograms use an array of counters [5] and array of accumulators with memory [2, 8]. In Memory based architectures, every pixel histogram computation needs one read and write memory access before and after computation. This kind of memory access time dominates the actual computation which eventually increases the processing time.

Li et al. [2] proposed a 256 × 16-bit memory array to store the number of pixels of each grayscale-level value. The grayscale level of the input pixel is used as the memory array address to read the pixel count from the relevant memory address, add 1 to it and write the updated count back to memory. In addition, three registers and an adder are required to read and update a particular grayscale level's count. The proposed design consumes almost 3 clock cycles to update the pixel count of a grayscale level..

Alsuwailem et al. [5] has proposed an architecture that uses a hierarchical decoder with an array of counters for simultaneous computation of histogram statistics and cumulative histogram statistics for all the 256 grayscale-level bins in the range 0 to 255. However, the cascaded logical OR gates at each output of hierarchical decoder add an extra propagation delay at the higher-order decoder outputs. The delay time for enabling the counters is increased by this delay at the decoder's higher order outputs. The maximum clock frequency that can be used as a system clock is limited by this increased delay period.

Nitin et al. [6] proposed architecture similar to Alsuwailem et al. [5] by realising the hierarchical decoder output equations with a set of logic gates to minimize the worst-case propagation delay for counters. However, this architecture leads to increased utilisation of hardware resources.

Soma et al. [11] proposed a hardware implementation approach of histogram equalisation algorithm on ZYNQ and KINTEX FPGA boards by utilising the IP cores generated in Vivado high-level synthesis (HLS) tool. Author concludes that the architecture implementation on ZYNQ FPGA board utilises less hardare resources compared to KINTEX FPGA Board.

Pulak Modal et al. [9], and Amritha Varshini et al. [12] have proposed fast VLSI architectures for parallel histogram computation. The architecture for computing the histogram is the only part of the proposed work; however, it does not address the architecture for computing the histogram equalisation.

Recent research work proposed by Younis et al. [10] and Sadad et al. [13] for implementation of histogram computation architecture uses a finite state machine-based approach. Proposed work has taken less space in FPGA by reducing the percentage of device utilisation at the cost of increase in computation time.

This research paper aims to propose a simple architecture to compute the histogram equalised image at a higher speed with moderate resource utilization. The proposed architecture uses 256 comparators to simultaneously compare the input pixel's grayscale level with all the grayscale levels in the range of 0 to 255 and 256 counters to compute the cumulative histogram. The counters are designed to increment their count value through combinational logic in advance through an enable signal from the comparator and update the output of the counter with the incremented value upon the arrival of the clock.

3 Proposed Work

The implementation of the histogram equalisation algorithm necessitates counting the number of occurrences of pixel grayscale levels ranging from 0 to 255 in an image. This paper proposes a novel high speed reconfigurable architecture to implement histogram equalisation algorithm on FPGA. The proposed architecture aims to compute an image's cumulative histogram by simultaneously updating the count of the current input pixel grayscale value and the count of all pixels whose values are greater than the current input pixel value. The cumulative histogram values are used to generate the final histogram equalised output values. Figure 1 demonstrates a top-level block representation of the proposed architecture for hardware realisation of histogram equalisation. Two input signals Pixel_in [7:0], clk and two output signals pixel_out [7:0], data valid are used by the top module. The signal description of the proposed histogram equalisation module is given in Table 1.

Table 1. Signal Description of Histogram Equalisation Module

Signals	Description
Pixel_in [7:0]	8 bit input Pixel grayscale level value
clk	System Clock Signal
Pixel_out [7:0]	Histogram equalised 8 bit output Pixel grayscale level value
Data_valid	Signal to indicate a valid output pixel grayscale level value

The proposed top module architecture comprises of comparators, counters and memory register array as shown in Fig. 2a. A detailed architecture for the proposed design comprising of 256 comparators, 256 counters and a 256 byte memory register array is shown in Fig. 2b. The design employs 256 comparators to compare the input pixel's grayscale level to each of the 256 grayscale levels in 0 to 255 range simultaneously and to produce a series of enable signals that enable the associated counters. The count of all the 256 grayscale values is updated using 256 counters. The parallel computing of the histogram and cumulative histogram for all pixel grayscale value in the range of 0 to 255 is made possible by an efficient design of comparators and counters. An array of 256 eight-bit registers is used to store the equalised output pixel values.

The proposed system uses a 64KB external RAM to store a 256 × 256 grayscale image. In each clock cycle, image pixel values are read one by one from external memory and compared simultaneously with all 256 grayscale levels ranging from 0 to 255 using 256 comparators. The design of comparators and counters ensures that the cumulative histogram for the input pixel value is updated in a single clock cycle.

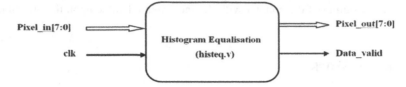

Fig. 1. Top module of Histogram Equalisation

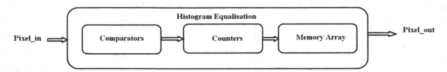

Fig. 2a. Architecture for Computing Cumulative Histogram Statistics

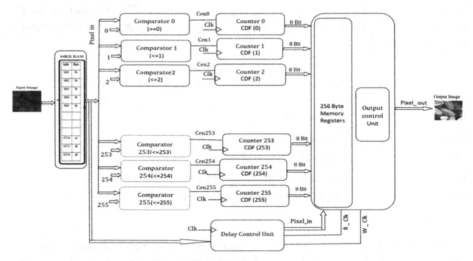

Fig. 2b. Detailed Proposed Architecture for Histogram Equalisation

3.1 Comparator Design

The proposed architecture for comparators has two 8-bit inputs (A,B) and one single-bit output (Y). Input A of the comparator receives the 8-bit input pixel, and Input B receives the 8-bit grayscale values in the range of 0 to 255. The output signal Y generates a logic 1 for condition A <= B. This output signal acts as a control signal to enable the associated counter to simultaneously update the histogram and cumulative histogram corresponding to the input pixel value. The proposed design employs 256 comparators to compare each grayscale level in the range of 0 to 255 with the input pixel in order to provide the enable signal for the corresponding 256 counters. This comparator design in the proposed work replaces the use of hierarchical decoder in the architecture proposed by [5]. Figures 3a and 3b show the block diagram and logic circuit of an 8-bit comparator used in the proposed design. An array of 8 two-input XNOR gates is used to check the equality condition of inputs A and B, and an array of 8 two-input AND gates is used to check the less than condition.

This proposed comparator design significantly contributes to the reduction of cumulative distribution function computation time (CDF).

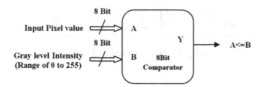

Fig. 3a. Block Diagram of 8 bit Comparator

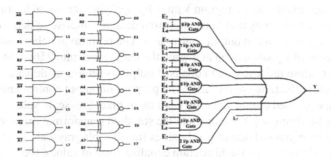

Fig. 3b. Logic Diagram of 8 bit Comparator

3.2 Counter Circuit Design

The counter architecture proposed in this research paper makes use of three input signals: reset, enable, and clock. Each counter in the design is responsible for counting the occurrences of the pixel grayscale values, which range from 0 to 255, while simultaneously updating the statistics of all pixel grayscale levels which are greater than the current input pixel grayscale value. The final histogram equalised output values are calculated directly from this accumulation process outputs. The block diagram and signal description for the proposed counter-design architecture is presented in Fig. 4.

Adv_cnt and res_cnt are the asynchronous signals used in the proposed counter design. The counter is reset using the signal res_cnt whenever the counter output reaches its maximum count. Adv_cnt signal of the counter gets its input from the output of the comparator. The counter increases its count value when the comparator output sets the Adv_ cnt signal to logic 1. The counter updates the output count with its incremented value for each clock cycle of the clock signal clk,.

All 256 counters used in the proposed architecture are designed to hold the cumulative histogram value of all the pixel values after N^2 clock cycles for an $N \times N$ image.

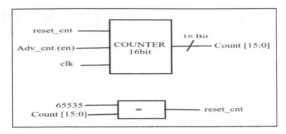

Fig. 4. 16 bit Counter Design

3.3   Pixel Mapping Unit Design

The main function of the pixel mapping unit is to compute the histogram equalised output values from the cumulative histogram values by applying the suitable transformation function. The pixel mapping unit in the proposed design comprises a delay control unit, 256 eight-bit registers, and an output control unit. The purpose of delay control unit is to generate the control signals to read and write the data values into 256 eight-bit registers that are used to store the histogram equalised output pixel grayscale value for all 8-bit grayscale levels in the range of 0 to 255. The purpose of output control unit is to output the histogram equalised output value for each input pixel grayscale value.

Equations (4) through (5) demonstrate the standard transformation function used to obtain histogram equalised output pixel values for all 8-bit grayscale levels. To reduce the time required to compute the histogram equalised output values, the proposed architecture uses the transformation function shown in Eq. (6). According to Eq. (6), the histogram equalised output pixel values are the most significant 8 bits of the 16-bit counter output.

By replacing the division operation in the standard pixel transformation function with a right shift operation, the most significant 8 bits of 16 bit counter output is considered as the histogram equalised output pixel value in the proposed design. This design feature reduces the overall computing time of the histogram equlisation to the greatest extent achievable.

$$S_k = (X_{L-1} - X_0) \times CDF(k) \text{ for } k = 0, 1, 2, 3, \ldots\ldots 255 \tag{4}$$

where S_k refers to histogram equalised output pixel intensity for the input pixel intensity k. Substituting Eq. (4) in Eq. (5) with N as 256×256, X_{L-1} as 255 and X_0 as 0, we get

$$Sk = \frac{255 \times \sum_{r=0}^{k} n_r}{256 \times 256} \text{ for } k = 0, 1, 2, 3 \ldots\ldots 255 \tag{5}$$

The simplified form of Eq. (5) is given in Eq. (6).

$$Sk = \frac{\sum_{r=0}^{k} n_r}{256} \text{ for } k = 0, 1, 2, 3 \ldots\ldots 255 \tag{6}$$

The time at which the most significant 8 bits of all 256 counter's output has to be stored in 256 registers is controlled by the delay control unit.

3.3.1 Delay Control Unit Design

Delay control unit is used to generate the control signals to store the computed histogram equalised pixel values in registers to map the input pixel with its corresponding transformed output pixel stored in 256 byte register array. The block diagram and detailed architecture of delay control unit is shown in Fig. 5. Delay control unit generates two control signals namely Write_En and Rd_clk.

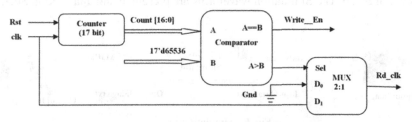

Fig. 5. Delay Control Unit Architecture

The computed histogram equalised output pixel value is stored into a 256 × 8 register array using the Write_En signal. For every input pixel read from the external image memory, an additional delayed clock signal Rd_clk was employed as a memory read control signal to read the histogram of equalised output pixels from a 256 × 8 register array. After N^2 input clock cycles, the delay control unit generates a read clock signal (Rd_clk) that is identical to the input clock signal (clk) and sets the Write_En signal to logic 1.

3.3.2 Output Control Unit

The output control unit is responsible to output the histogram equalised pixel value for each input pixel. This unit employs a multiplexer to read the histogram equalised pixel value for each input pixel value in every cycle of Rd_clk.

4 Experimental Results

Proposed histogram equalisation architecture has been coded in matlab to ensure the correctness of the developed algorithm. Subsequently the proposed architecture has been developed using RTL compliant verilog HDL (Hardware Description Language) code so that it can be implemented on an FPGA.

A Matlab code has been written to read the pixel intensities of a standard image (hexadecimal value) in.jpg and.bmp format and store it in a text (.txt format) file. Different variety of images such as low contrast dark, bright, medium contrast and medical images from different datasets [14, 15] are used to test the functionality of the proposed architecture. Grayscale scale and RGB color images of size 256 × 256 and 512 × 512 have been used in simulation. The text file generated by the Matlab code has been used by Xilinx ISIM (Ver. SE 14.7) for simulation.

A top-level test bench module has been developed in verilog to pass the pixel intensities as a stimulus for the top design histogram equalisation module to generate the outputs of the top level design module (histeq.v). A developed test bench module has been run using Xilinx ISIM (Version 14.7) to write the output in an output text (.txt format) file. To display the histogram-equalized output image, a Matlab program has been written to read the pixel intensities from the output text file and convert them back into an output image in the.jpg (.jpg format) file. Entire coding and simulation procedure is shown in Fig. 6. The Simulation waveforms are presented and analysed in Sect. 4.1

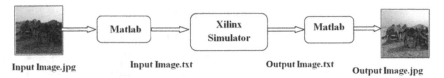

Input Image.jpg   Input Image.txt   Output Image.txt   Output Image.jpg

Fig. 6. Simulation Process

4.1 Simulation Results

The simulation results of proposed reconfigurable architecture for histogram equalisation on a image size of 256×256 are presented in Fig. 7. A clock signal with a period of 10ns has been used as the system clock for simulation. It is evident from simulation results that the proposed architecture of histogram computation unit reads one pixel in one clock period and generates the histogram equalised output pixels for all 65,536 pixels of 256 $\times$ 256 image size in 6,55,360ns. After 6,55,360 ns, the design generates a data valid signal dvm indicating that the output bus Pixel_out[7:0] has a valid histogram equalised output pixel value for a given input pixel of the image.

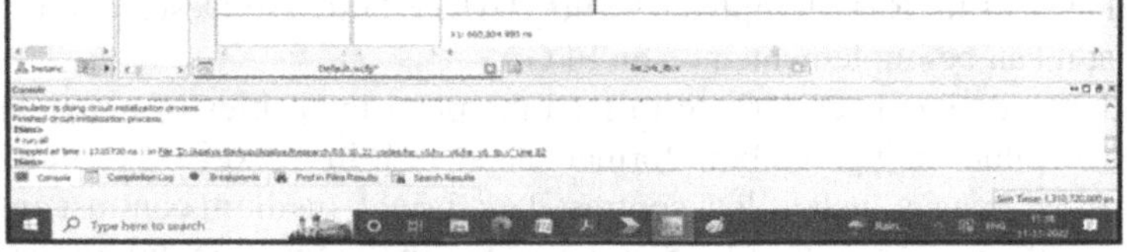

Fig. 7. Simulation Output Waveform - Output Pixels Values for 220th to 225th Input Pixels

4.2 Synthesis Results and Analysis of Proposed Architecture

The XILINX ISIM verified RTL code has been synthesized and implemented on Kintex 7 low voltage FPGA family device Xc7k70tl using Xilinx XST synthesis tool in ISE Project navigator (Ver 14.7). The proposed design is implemented on to the target FPGA device and routed to various configurable logic blocks, slices and LUT's. Figure 8 shows the RTL schematic of proposed histogram equalisation top module with all its integrated sub blocks like comparators, counters, register array and delay control unit. From the detailed RTL schematic, it is evident that the sub blocks and signals in the implemented design matches the architecture blocks and signals proposed in Sect. 4.

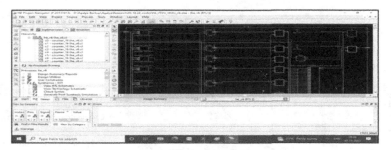

Fig. 8. Detailed RTL Schematic

The significance of the proposed architecture for histogram equalisation algorithm has been portrayed using the percentage of device utilisation and the computation time. The summary of percentage of device utilisation in terms of slice registers, LUT's and IOB's for implementing the proposed design on Xilinx Kintex 7 FPGA device Xc7k70tl for different image size's 256 × 256 and 512 × 512 is presented in Fig. 9. From the device utilisation summary chart it is evident that the proposed design utilizes only 2% more of slice registers and LUTs to implement the design for an image size of 512 × 512 while compared to 256 × 256 image size without any significant additional cost of device utilisation. So it is found that the proposed design is more suitable to support VGA and SVGA standard image sizes without much additional cost of device utilisation.

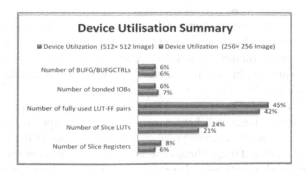

Fig. 9. Device Utilization Summary for 256 × 256 and 512 × 512 image size

The timing report generated after implementing the proposed design on Kintex 7 FPGA device reveals that the maximum frequency of operation of the proposed design is 425 MHz with a processing time of 2.353 ns.

The computaion time of the proposed histogram equalisation architecture has been compared with computation time of the architectures proposed [5, 7, 11, 14]. From the results shown in Fig. 10 it is found that the proposed architecture design computes the histogram equalised output pixel values in very much lesser time compared to other existing design architecture proposed by [5, 7, 11, 14]. This increase in computation speed creates new possibilities for the proposed design, by making it more appropriate for real-time applications.

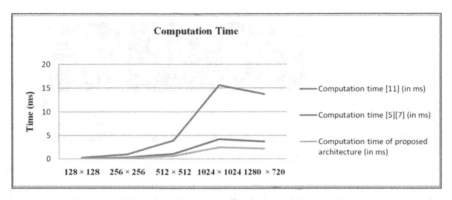

Fig. 10. Computation Time – Comparison

A variety of images, including low contrast dark, bright, medium contrast, and medical images from various datasets [14, 15] have been used to validate and test the FPGA implementation of the proposed architecture for histogram equalisation and compared with matlab implementation. Although more than a dozen of both grayscale scale and RGB color images in.jpg and.bmp format with a resolution of 256×256 and 512×512 images have been used to test the proposed architecture, a set of sample images with its output is shown in Fig. 11. From the results, it is found that the visual appearance and histogram plot of output images obtained from matlab implementation and FPGA implementation looks similar.

In Fig. 11 original low contrast grayscale input images, output images obtained from proposed histogram equalisation architecture implemented using matlab and verilog HDL are presented in column1, 2 and 3 respectively. The histogram spread of input images and output images obtained from matlab and FPGA implementation are presented in column 4, 5 and 6 of Fig. 11.

The level of histogram spread over all the bins of image signifies the contrast of image. Low contrast images will have a narrow histogram spread and high contrast images will have a wider histogram spread over the entire bins of the image. From the histogram plot of input and output images shown in Fig. 11, it is evident that the output images have improved brightness and contrast compared to original input images.

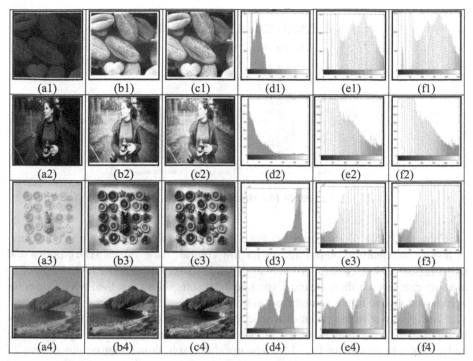

Fig. 11. (a1, a2, a3, a4). Input image, (b1, b2, b3, b4): Histogram equalized output image using matlab (c1, c2, c3, c4): Histogram equalized output image using verilog, (d1, d2, d3, d4): Input image Histogram (e1, e2, e3, e4): Histogram of output image using matlab, (b3, d3, f3, h3): Histogram of output image using verilog.

The common qualitative metrics like Peak Signal to Ratio (PSNR) and entropy presented in Eq. (7) and (9) has been used to measure the performance of the proposed architecture implemented using verilog and matlab coding. A comparative chart of measured PSNR and entropy for images having different levels of contrast is shown in Figs. 12 and 13.

$$PSNR = 10 \times \log_{10}(255 \times 255/MSE) \tag{7}$$

$$MSE = \frac{1}{m \times n} \sum_{x=0}^{m} \sum_{y=0}^{n} (I(x, y) - O(x, y))^2 \tag{8}$$

MSE – Mean Square Error, I(x, y) - Input Image, O(x, y) – Output Image

$$E = -\sum_{i=0}^{n-1} P_i \, log_2 \, P_i \tag{9}$$

where E refers to Entropy of image, n = 256 for 8 bit grayscale levels Pi refers to probability of pixels with grayscale level i.

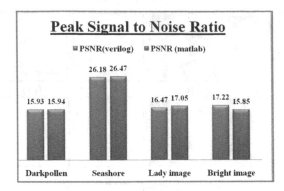

Fig. 12. PSNR Plot Histogram Equalised Images

From the PSNR plot presented in Fig. 12, it is found that Peak signal to noise ratio of histogram equalised output images for low contrast dark images (Dark pollen and lady image) and low contrast bright image (Bright image) is found be less compared to medium contrast sea shore image. Experimental results shows that the PSNR values for verilog and matlab implemented are almost same.

Entropy is one of the common metric to measure the quality of an image. Entropy of input and output images of the proposed architecture implemented using verilog and matlab are plotted in Fig. 13. Experimental result shows that output images of low contrast input images (Dark pollen, Lady image and Bright image) have improved entropy than medium contrast input image (Sea shore image).

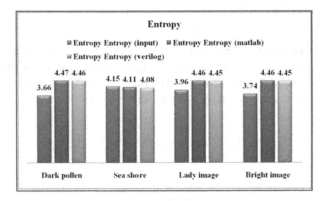

Fig. 13. Entropy of input and output images of proposed design

5   Conclusion

The primary objective of this research work is to develop reconfigurable hardware architecture for histogram equalisation that requires less computation time and moderate hardware resources. The proposed architecture has been coded in both matlab and RTL compliant verilog HDL and verified its functionality through simulation using Xilinx integrated simulator tool Xilinx ISIM. The Proposed design has been synthesized and implemented on Xilinx Kintex 7 low voltage FPGA family device Xc7k70tl. From the experimental results, it is found that the computation speed of the proposed architecture for implementing histogram equalisation on FPGA is 1.8 times faster than the existing architectures proposed in [5, 6] and more than 6 times faster compared to the architectures proposed in [10, 13]. This increase in computation speed presents new possibilities for improving the appropriateness of the proposed design for real-time applications.

References

1. Kim, Y.T: Contrast enhancement using brightness preserving bi - histogram equalisation. IEEE Trans. Consum. Electron. **43**(1), 1–8 (1997)
2. Li, X., Ni, G., Cui, Y., Pu, T., Zhong, Y.: Real-time image histogram equalization using FPGA. Elec. Imaging Multimedia Syst. II **3561**, 293–299 (1998). https://doi.org/10.1117/12.319719
3. Wang, E.Y., Chen, Q., Zhang, B.: Image enhancement based on equal area dualistic sub-image histogram Equalisation method. IEEE Trans. Consum. Electron. **45**(1), 68–75 (1999)
4. Chen, S.-D., Ramli, A.R.: Minimum mean brightness error bi-histogram equalization in contrast enhancement. IEEE Trans. Consum. Electron. **49**(4) 1310–1319 (2003)
5. Alsuwailem, A.M., Alshebeili, S.A.: A new approach for real-time histogram equalization using FPGA. In: International Symposium on Intelligent Signal Processing and Communication Systems, pp. 397–400 (2005). https://doi.org/10.1109/ISPACS.2005.1595430
6. Sachdeva, N., Sachdeva, T.: An FPGA based real-time histogram equalization circuit for image enhancement. IJECT **1**(1), December 2010
7. Agalya, P., Hanumantharaju, M.C., Gopalakrishna, M.T.: A detailed review of color image contrast enhancement techniques for real time application. In: Information Systems Design and Intelligent Applications, pp. 487–497 (2016).https://doi.org/10.1007/978-81-322-2757-1_48
8. Jiao, H., Xing, J., Zhou, W.: Histogram equalization image enhancement based on FPGA algorithm design and implementation. In: Hung, J., Yen, N., Hui, L. (eds.) in Frontier Computing. FC 2018. Lecture Notes in Electrical Engineering, vol. 542. Springer, Singapore (2019). https://doi.org/10.1007/978-981-13-3648-5_173
9. Mondal, P., Banerjee, S.: A reconfigurable memory-based fast VLSI architecture for computation of the histogram. IEEE Trans. Consum. Electron. **65**(2), 128–133 (2019). https://doi.org/10.1109/TCE.2019.2900541
10. Younis, D.B., Younis, B.M.: Low cost histogram implementation for image processing using FPGA. IOP Conf. Ser. Mater. Sci. Eng. **745**(1), 012044. IOP Publishing, February 2020
11. Soma, P., Sravanthi, C., Srilakshmi, P., Jatoth, R.K.: Implementation of single image histogram equalization and contrast enhancement on Zynq FPGA. In: Laxminidhi, T., Singhai, J., Patri, S.R., Mani, V.V. (eds.) Advances in Communications, Signal Processing, and VLSI. Lecture Notes in Electrical Engineering, vol. 722. Springer, Singapore (2021).https://doi.org/10.1007/978-981-33-4058-9_7
12. Varshini, R.A.: A reconfigurable memory based fast VLSI architecture for histogram computation. Turkish J. Comput. Math. Educ. (TURCOMAT) **12**(6), 244–252 (2021)
13. Ul Sadad, N., Afrin, A., Mondal, M.N.I.: FPGA based histogram equalization for image processing. In: 2021 3rd International Conference on Electrical and Electronic Engineering (ICEEE), Rajshahi, Bangladesh, pp. 89–92 (2021). https://doi.org/10.1109/ICEEE54059.2021.9718837

14. Qureshi, M.A., Sdiri, B., Deriche, M., Faouzi, A.-C., Beghdadi, A.: Contrast Enhancement Evaluation Database (CEED2016). Mendeley Data, V3 (2017). https://doi.org/10.17632/3hf zp6vwkm.3
15. https://medpix.nlm.nih.gov/casebydiagnosis

A Practical Use for AI-Generated Images

Alden Boby[✉] , Dane Brown , and James Connan

Department of Computer Science, Rhodes University, Grahamstown, South Africa
boby.alden128@gmail.com, {d.brown,j.connan}@ru.ac.za

Abstract. Collecting data for research can be costly and time-consuming, and available methods to speed up the process are limited. This research paper compares real data and AI-generated images for training an object detection model. The study aimed to assess how the utilisation of AI-generated images influences the performance of an object detection model. The study used a popular object detection model, YOLO, and trained it on a dataset with real car images as well as a synthetic dataset generated with a state-of-the-art diffusion model. The results showed that while the model trained on real data performed better on real-world images, the model trained on AI-generated images, in some cases, showed improved performance on certain images and was good enough to function as a licence plate detector on its own. The study highlights the potential of using AI-generated images for data augmentation in object detection models and sheds light on the trade-off between real and synthetic data in the training process. The findings of this study can inform future research in object detection and help practitioners make informed decisions when choosing between real and synthetic data for training object detection models.

Keywords: Diffusion Probabilistic Models · Image Synthesis · Licence Plate Recognition · Object Detection

1 Introduction

Collecting data for research purposes can be time-consuming for multiple reasons. It may require ethics clearance, as well as travel to specialised areas for data collection in the field [9]. Research could greatly benefit from the use of artificial data close enough to reality to be used practically for training machine learning models, especially for areas lacking such data. Collecting and labelling data can account for up to 90% of the time in research experiments, greatly limiting the rate at which new technologies can advance [23].

Often machine learning models perform much better based on the amount of data they have access to, as well as some variation in the data to allow the model to be well-rounded [23]. Collecting large datasets can take years, delaying the time it takes to make advancements. This constraint can be reduced through some means of automatically generating data; with some already assisted data labelling algorithms, data collection time could be cut severely.

J. Abawajy et al. (Eds.): ICICCT 2023, CCIS 1841, pp. 157–168, 2023.
https://doi.org/10.1007/978-3-031-43838-7_12

Recently, images generated through artificial intelligence have gained popularity and have received lots of attention due to how visually impressive they are and how closely they resemble reality [15, 16]. This raises the question of how well a machine learning model trained on purely synthetic data would work when applied in the real world. A successful model would qualify as a means of reducing time related to data collection, reducing costs and speeding up research. This paper intends to investigate this viability by using synthetic data to train an object detection model to detect licence plates on cars. With many public datasets available, vehicle-based data is generally easier to access; vehicles are a good way to test this hypothesis. Many of these public datasets lack diversity and cater specifically to one country. The images often display rear and frontal views and lack more challenging detection angles, leading to reduced performance in unconstrained scenarios [17]. There is enough vehicle data to create convincing images of vehicles. The experiments' findings will provide insight into how useful these images are and if they provide utility beyond content creation.

2   Diffusion Probabilistic Model

Diffusion probabilistic models create images from Gaussian noise, gradually denoising an image until the result is present. This is combined by training a model using a frozen Contrastive Language-Image Pre-training (CLIP) text feature extractor to map words to objects so that suitable images can be generated with prompts from an end user [15]. CLIP allows the model to learn to associate certain words or phrases with specific features or attributes in the images, which can then be used to control the generation process and produce images with specific desired characteristics. The training of the model utilised the LAION 5B dataset, which comprises a massive collection of 5.85 billion image-text pairs that have been filtered using CLIP [18].

This diffusion model for image synthesis is particularly useful for generating realistic images, as it considers the natural statistical properties of images, such as the correlation between neighbouring pixels. Using a Markov chain series, the model can also ensure that the final image is coherent and visually pleasing [6].

The likelihood of each step is solely determined by the state achieved in the preceding event. As the iterations progress, the image gradually becomes more coherent, with similar pixel values appearing in regions of the image where they are likely to be found [3]. Eventually, the image reaches a steady state where it is unlikely to change further. Figure 1 shows a visual representation of the process.

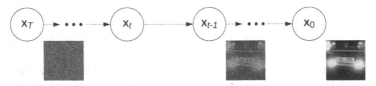

Fig. 1. A visual representation of how diffusion models create an image via the Markov Chain, gradually reversing the noise in an image, x_t, x_{t-1},..., until a clear image is produced at x_0.

Using a latent diffusion model, the output data is in control of the researchers, for example, it can be used to create an image of a car in snow or dust, allowing for a license plate detector to be trained to work in differing environments while escaping the need to collect the data from such environments physically. A disadvantage of these models is that they are not particularly great when it comes to generating text, so currently, the benefits of this technology can only be realised for the initial detection of licence plates. The model has some potential to be used for vehicle detection. However, this requires the generated images to be very accurate as the features of a car model are consistent, and the model tends to produce images with slightly different features each time.

These models work with a linear time complexity, as the number of iterations grows, there is a need for a commensurate increase in the amount of time needed, this, in particular, does not impact data collection times as these models are still running in a reasonable time, minutes, as opposed to days when compared to collecting suitable real-world data. As the model only generates images, the ground truth bounding boxes will still have to be labelled manually for training. The model allows for batches of images to be generated reducing the time spent making prompts.

3 Object Detection

YOLO has proven to be one of the best object detectors and can easily be modified to detect a range of objects subject to its training data [21]. The model's speed of inference and minimal accuracy trade-off makes it the more favoured option compared to other existing models like the Region-based Convolutional Neural Network (R-CNN), and faster RCNN [5]. YOLO is a single-shot object detector which allows it to classify an object in one pass, making it capable of performing in real-time; this is where it gains its advantage over RCNN and its variants which are two-stage detectors [11].

The YOLO model at high-level works by splitting an image into an n x n grid. Suppose the centre of an object is present in a cell of the grid that cell is classified as that object. From there, each block is assigned a confidence score, indicating the likelihood that a specific object matches the model's prediction. YOLO can assign multiple bounding boxes in a cell, since the same instance of an object may be detected several times, this creates overlapping bounding boxes, which are undesirable, non-maximum suppression is used to remove redundant bounding boxes resulting in one bounding box for an object [14].

4 Related Studies

In recent years, object detectors such as YOLO have become prominent in the licence plate detection space, with many existing implementations making use of them and often being compared against other object detectors to find the best match. Lee *et al.* [10] made use of Faster-RCNN as well as YOLO and concluded that YOLO was the superior model to use when it came to licence plate detection, based on its speed of inference, it was found that the accuracy improvements Faster-RCNN had were not enough to justify the use of that architecture over YOLO. Boby *et al.* [2] also made use of YOLO for licence plate recognition specifically by training a YOLO model by gathering training data representing challenging scenarios in order to make a robust licence plate detector. A variation of the YOLO model created by Silvia and Jung, the warped planar object detection network (WPOD-net) [19], crafted specifically for the licence plate domain and can be considered state-of-the-art as it is highly accurate and has one main advantage over existing models. A unique contribution from their model is bounding parallelograms which are a much better way of encapsulating the region of interest as they are invariant to oblique angles in a frame or image, allowing the model to represent the shape of licence plates accurately regardless of the placement of the licence plate.

Generative Adversarial Networks have led in the area of image synthesis [24], with popular models such as StyleGAN, which can generate unique faces [7]. Using GANs to generate images of vehicles has been done before. Unfortunately, the results were less than satisfactory when viewed solely from a visual fidelity standpoint. Results from Kim [8] regularly produced images resembling vehicles but lacked certain features such as wheels and lower-level features such as a licence plate. Images from a GAN are expected to have a wide variety. GANs are specifically hard to train, the model can be affected by a phenomenon known as mode collapse, causing the model to produce similar-looking images every time [20]. This is undesirable, especially when trying to create diverse images.

Generative Adversarial Networks were the preferred choice prior to advancements regarding diffusion models. Dhariwal et al. [4] elaborate on how the newer diffusion models produce much more satisfying results in the image synthesis space. A few diffusion models have gained popularity in the past year, creating realistic images that can be hard to distinguish from real life without a trained eye. These models have been used extensively on social media, and for content creation [15, 16]. Given that these models are relatively recent, there has been little to no application in specific fields, and there lack many practical uses for them other than in media and the art community. In addition to producing more realistic output, latent diffusion models are much easier to train than GANs [4]. Moreover, diffusion models have been proven to produce higher quality super-resolution images than super-resolution-based GANS such as the SRGAN, which has been popular in recent times. Both models overlap in their functions, and the promising results from diffusion models may see GANs being replaced.

Perceiving the quality of an image is difficult as there is no direct metric that correlates directly to how humans perceive an image [22]. Quantitative metrics have been developed in an attempt to quantify image quality, such as SSIM and PSNR however, quantifying the quality of a generated image from AI models is challenging due to the fact that all existing performance metrics for judging image quality require a reference image to be compared to. As such the best way to examine generated images would be through visual inspection.

Batches of images can be generated for each prompt, which creates a large volume of images which can quickly be sorted through and selected as desired, presenting the proposed speed up from synthesised data. Licence plate detection presents a safe and low-risk environment to test the effectiveness of synthesised data, as opposed to medical fields, which make use of computer vision [12].

5 Proposed Work

This paper uses a stable-diffusion model trained on the LAION-5B, which contains 5.85 billion images paired with descriptive text used to give context to the images, the dataset includes vehicle-related data making it suitable for generating images of vehicles [18]. Using the separate datasets, two YOLOv7 models will be fine-tuned for licence plate detection. One model will undergo training exclusively using synthetic images and undergo testing using real-world images, while a separate identical model will be trained and tested on real-world data. The model's performance will be measured by comparing the performance of the real-world data based model and the synthetic data based model. These models will be trained in isolation, meaning they will both have separate training data but will be trained with the same parameters.

Output from the diffusion model will be analysed through visual inspection to show how well the model can generate images based on prompts given to it by a user. Given that not all images from the diffusion model are of good enough quality to be used, some human reasoning is required to discern whether or not an image is good enough to be added to the dataset for training.

5.1 Datasets

The proposed licence plate detection systems were evaluated and trained on public datasets besides the newly created Stable Cars dataset. To ensure consistency with the input size of the YOLOv7 model, the training data was resized to 640 × 640.

Stable Cars
The Stable Cars dataset, named after the model used to generate them, Stable Diffusion [15], consists of 100 images of vehicles completely synthetic images of cars. The images in the dataset have varying weather conditions to demonstrate the flexibility of image generation.

MediaLAB Licence Plate Recognition Dataset
A collection of 433 car images with varied lighting conditions and inconsistent camera angles makes it a good dataset to work with for testing, and validation [1].

Croatian Licence Plate Dataset
The Croatian Licence plate dataset is a collection of vehicles captured from the rear end featuring a variety of lighting conditions such as sunny, cloudy and lowlight (nighttime). It consists of over 500 images.

5.2 Performance Metrics

Mean average precision, IoU, recall, precision were used to evaluate the performance of the models. These metrics have been established for quantitative evaluation of the performance of object detection models, are useful for comparing different models and are standard throughout literature [11].

Intersection Over Union
This measures the overlap between the predicted bounding box and the ground truth bounding box, it is expressed as a ratio between the area of overlap and the area of union. The equation for IoU is shown in Equation 1.

$$\text{Intersection over Union} = \frac{\text{Area of Overlap}}{\text{Area of Union}} \tag{1}$$

Mean Average Precision
The Mean Average Precision calculates the average precision by considering all classes, for licence plate detection, there is only one class. mAP summarises a model's ability to detect objects of various classes within an image, considering both the accuracy of the detections and their consistency across different images. A higher mAP correlates to better model performance.

Precision
Precision measures the accuracy of the detections, indicating what proportion of the detections made by a model were true. A high precision value indicates a low false positive (FP) rate for the model, while a low precision value means that many of the detections made by the model were not of the correct class. The formula for precision is defined in Equation 2.

$$\text{Precision} = \frac{TP}{TP + FP} \tag{2}$$

Recall
Recall can be described as the total of correctly identified positives (TP), divided by the sum of true positives and false negatives (FN). The formula for recall is defined in Equation 3. Recall measures how many objects in supplied data have been correctly detected by the model. A high recall value indicates that the model is good at finding the desired classes. In contrast, a low recall value indicates that the model cannot detect the majority of instances in the data.

$$\text{Recall} = \frac{TP}{TP + FN} \tag{3}$$

F1 Score
The F1 score is a value that can be used to find the best balance between precision and recall. The F1 score indicates the optimum confidence value to be used when evaluating data. The optimal confidence value will depend on the model's performance when it comes to precision and recall. The formula for the F1 score is defined in Equation 4.

$$F_1 = 2 \cdot \frac{\text{precision} \cdot \text{recall}}{\text{precision} + \text{recall}} \tag{4}$$

5.3 Model Training

Two separate YOLOv7 models were trained, the first model was trained with the Stable Cars dataset, which consists of artificial images of cars. The second model was trained on the MediaLAB Licence plate dataset and the Croatian licence plate dataset. All datasets had to be manually labelled, including the synthesised dataset. The training split model was 100 training images, 25 validation images and 100 test images. From the split, the same validation and test images were used for the two models while they were trained with different data. The Stable Cars dataset was only involved in training the model. Both models were trained with a batch size of 8 for 300 epochs, using existing weights trained on the COCO dataset.

- CPU: Ryzen 3950X 4.7 GHz 16-Core
- GPU: 2080TI 11GB VRAM
- RAM: 128GB

6 Result Analysis

Generating the dataset of synthetic images required careful wording when crafting the prompts, for example, explicitly mentioning a licence plate in the prompt forces the licence plate to be merged with surroundings or put in strange places. Fig. 2 demonstrates an example of including a licence plate within the prompt.

The overall look of the image itself was stylised and not realistic. So the prompts used to generate images refrained from using licence plate as a keyword to produce more natural output, as with a car, a licence plate is implied, and the same is true with the training data, the licence plate is a feature of the vehicle and does not explicitly have to be declared and reduces the parameters the model needs to work with.

Fig. 2. "A car driving on a bridge with the licence plate showing".

Unlike images produced from a GAN [8], the vehicles generated for the Stable Cars dataset have distinct features as well as a licence plate which was noticeably harder to reproduce through the use of a GAN model trained to generate car images. Figures 3 and 4 provide visual representations of samples from the generated dataset, observation yields that employing an AI model to create a dataset can allow for the creation of a very diverse dataset, including both snowy and rainy weather, which is not easy to do in real life. Using brand names within the prompt significantly improved the frequency of licence plates appearing in the resulting image as well as helping the model to produce an overall more convincing image. In most cases, the images generated are sufficient to train a model to locate a licence plate's general location, but further use of them for character recognition is not currently feasible. Using this information the prompts used to generate the data included keywords such as 'snow' and 'rain' to alter the weather in the resultant images.

Fig. 3. The AI-generated image presents a difficult licence plate detection scenario because of its positioning in the image.

Table 1 summarises the performance of both YOLO-based licence plate detection models.

Overall the model trained on real-world data performed better across the board, however, the model trained on synthetic data is not far behind and produced a useable model that could make predictions on real data. Table 1 cannot fully justify recall and precision as they are better observed over a series of confidence values. Fig. 5a and b show the synthetic and real model recall curves, respectively.

Looking at the recall graphs, it can be seen that both models have recall values above 70% up until the confidence value reaches about 0.9. The recall graph for the synthetic model Fig. 5a is not as smooth as the opposite with significant drops in recall with increased confidence as opposed to the relatively smooth drop in recall for the

Fig. 4. Having snow in training data would previously require data collection in specific areas as it does not snow in every region.

Table 1. High level comparison of both YOLOv7 models.

Model	mAP@0.5	Recall	Precision
YOLOv7 (Synthetic)	0.809	0.96	1.00
YOLOv7 (Real)	0.980	0.98	1.00

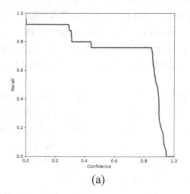

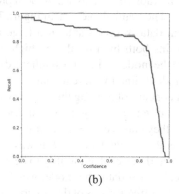

(a)   (b)

Fig. 5. (a) Recall curve for the YOLO model trained on AI-generated images. (b) Recall curve for the YOLO model trained on real-world data.

model trained on real data. The synthetic model suddenly drops in recall between the confidence values 0.3 and 0.5. This shows that the synthetic model is more prone to missing predictions than the model trained on real-world data.

The opposite is true for precision, a very high confidence will severely limit the number of predictions the model can make, meaning the model will be very precise but will miss out many licence plates unless it is very sure. Comparatively, the synthetic trained model does not perform as well when it comes to precision, as at lower confidence values, the precision is below 0.5, whereas for the model trained on real data, the precision is almost consistently at a value of 0.8. The erratic data points on the precision curve for the synthetic model demonstrate the synthetic model is more prone to false positives, especially at lower confidence levels (Fig. 6).

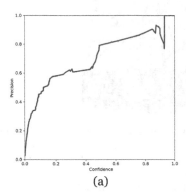 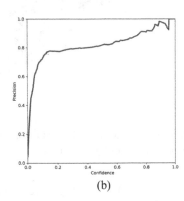

(a)   (b)

Fig. 6. (a) Precision curve for the YOLO model trained on AI-generated images. (b) Precision curve for the YOLO model trained on real-world data.

The F1 score is useful in determining the best confidence values to use when running the model. Figs. 7b and a show the graphs for the F1 curves for both models. The peak of each graph represents where both precision and recall are maximal for the respective model. The graph depictedT in Figure 7b demonstrates the model's performance trained on real data is very stable at a large range of confidence values meaning the model maintains both high recall and high precision at a wider range of confidence values. While the model trained on synthetic data is still useable, it cannot outperform or replace the model trained with real data.

A concern of training the models with synthetic is that licence plates in the generated images often have illegible text as these diffusion models have not yet reached a point where they can accurately generate text on command. As the YOLOv7 model achieved an overall mAP@0.5 of 809, it was found that the artefacts within the images did not significantly affect their ability to be used as training data, as even though they were not fed real-world data, the model was fully capable of making predictions on real-world data. The effectiveness of deep learning models in achieving their tasks relies a great deal on the quality as well as the quantity of data supplied to it. A use case for AI-generated data would be to supplement real-world data to increase the quantity of data available rather than relying solely on them.

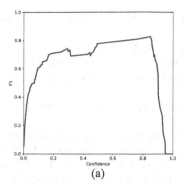
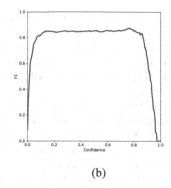

(a)   (b)

Fig. 7. (a) F1 score curve for the YOLO model trained on AI-generated images. (b) F1 score curve for the YOLO model trained on real-world data.

7 Conclusion

In conclusion, this research paper has compared the use of real data and AI-generated images for training an object detection model. The results showed that while the model trained on real data performed better on real-world images, the model trained on AI-generated images showed improved performance on images with variations such as rotations and scaling. The study highlights the potential of using AI-generated images for data augmentation in object detection models and sheds light on the trade-off between real data and synthetic data in the training process. These findings can inform future research in object detection and help practitioners make informed decisions when choosing between real and synthetic data for training object detection models. The findings indicate that employing a combination of real and synthetic data could potentially be the most effective strategy for training object detection models, as it leverages the strengths of both data sources to improve performance. Future work could expand this research and look into using AI-generated images to train vehicle detection models. There are still some advancements required to use synthetic data for vehicle tracking as the AI often mixes features of different brands, which is counteractive for vehicle detection.

References

1. MediaLab lpr dataset. medialab.ntua.gr/research/LPRdatabase.html. Accessed 07 Feb 2023
2. Boby, A., Brown, D.: Improving licence plate detection using generative adversarial networks. In: Iberian Conference on Pattern Recognition and Image Analysis, pp. 588–601. Springer (2022). https://doi.org/10.1007/978-3-031-04881-4_47
3. Cheng, S.I., Chen, Y.J., Chiu, W.C., Tseng, H.Y., Lee, H.Y.: Adaptively-realistic image generation from stroke and sketch with diffusion model. In: Proceedings of the IEEE/CVF Winter Conference on Applications of Computer Vision, pp. 4054–4062 (2023)
4. Dhariwal, P., Nichol, A.: Diffusion models beat gans on image synthesis. Adv. Neural. Inf. Process. Syst. **34**, 8780–8794 (2021)
5. Diwan, T., Anirudh, G., Tembhurne, J.V.: Object detection using yolo: challenges, architectural successors, datasets and applications. In: Multimedia Tools and Applications, pp. 1–33 (2022)

6. Ho, J., Jain, A., Abbeel, P.: Denoising diffusion probabilistic models. Adv. Neural. Inf. Process. Syst. **33**, 6840–6851 (2020)
7. Karras, T., Laine, S., Aila, T.: A style-based generator architecture for generative adversarial networks, pp. 4396–4405, June 2019. https://doi.org/10.1109/CVPR.2019.00453
8. Kim, D.H.: Deep convolutional gans for car image generation. arXiv preprint arXiv:2006. 14380 (2020)
9. Lee, I., Shin, Y.J.: Machine learning for enterprises: applications, algorithm selection, and challenges. Bus. Horiz. **63**(2), 157–170 (2020)
10. Lee, Y., Yun, J., Hong, Y., Lee, J., Jeon, M.: Accurate license plate recognition and super-resolution using a generative adversarial networks on traffic surveillance video. In: 2018 IEEE International Conference on Consumer Electronics-Asia (ICCE-Asia), pp. 1–4. IEEE (2018)
11. Miller, D., Moghadam, P., Cox, M., Wildie, M., Jurdak, R.: What's in the black box? The false negative mechanisms inside object detectors. IEEE Robot. Autom. Lett. **7**(3), 8510–8517 (2022)
12. Osokin, A., Chessel, A., Carazo Salas, R.E., Vaggi, F.: Gans for biological image synthesis. In: Proceedings of the IEEE International Conference on Computer Vision, pp. 2233–2242 (2017)
13. Redmon, J., Divvala, S.K., Girshick, R.B., Farhadi, A.: You only look once: Unified, real-time object detection. CoRR abs/1506.02640 (2015)
14. Rombach, R., Blattmann, A., Lorenz, D., Esser, P., Ommer, B.: High-resolution image synthesis with latent diffusion models. In: Proceedings of the IEEE/CVF Conference on Computer Vision and Pattern Recognition, pp. 10684–10695 (2022)
15. Saharia, C., et al.: Photorealistic text-to-image diffusion models with deep language understanding (2022)
16. Sanyal, R., Jethanandani, M., Reddy, G.D., Kurtakoti, A.: Localizing license plates in real time with retinanet object detector. In: Sharma, M.K., Dhaka, V.S., Perumal, T., Dey, N., Tavares, J.M.R.S. (eds.) Innovations in Computational Intelligence and Computer Vision. AISC, vol. 1189, pp. 570–577. Springer, Singapore (2021). https://doi.org/10.1007/978-981-15-6067-5_64
17. Schuhmann, C., et al.: Laion-5b: An open large-scale dataset for training next generation image-text models (2022)
18. Silva, S.M., Jung, C.R.: A flexible approach for automatic license plate recognition in unconstrained scenarios. IEEE Trans. Intell. Transport. Syst. **6**, 5693–5703 (2021)
19. Srivastava, A., Valkov, L., Russell, C., Gutmann, M.U., Sutton, C.: Veegan: reducing mode collapse in gans using implicit variational learning. Adv. Neural Inf. Process. Syst. **30** (2017)
20. Wang, C.Y., Bochkovskiy, A., Liao, H.Y.M.: Yolov7: Trainable bag-of-freebies sets new state-of-the-art for real-time object detectors. arXiv abs/2207.02696 (2022)
21. Wang, Z., Bovik, A.C., Lu, L.: Why is image quality assessment so difficult? In: 2002 IEEE International Conference on Acoustics, Speech, and Signal Processing, vol. 4, pp. IV–3313. IEEE (2002)
22. Whang, S.E., Lee, J.G.: Data collection and quality challenges for deep learning. Proc. VLDB Endow. **13**(12), 3429–3432 (2020). https://doi.org/10.14778/3415478.3415562
23. Wu, X., Xu, K., Hall, P.: A survey of image synthesis and editing with generative adversarial networks. Tsinghua Sci. Technol. **22**(6), 660–674 (2017)

Homomorphic Encryption Schemes Using Nested Matrices

Ashwaq Khalil⬤, Remah Younisse⬤, Ashraf Ahmad⬤, and Mohammad Azzeh$^{(\boxtimes)}$⬤

Department of Computer Science, Princess Sumaya University for Technology, Amman, Jordan
{ash20219002,rem20219007}@std.psut.edu.jo, {a.ahmad,
m.azzeh}@psut.edu.jo

Abstract. With the increasing necessity to secure data stored in different cloud platforms and transmitted through the internet, many encryption algorithms are developed to protect data from attacks. Homomorphic encryption (HE) which is used to allow performing arithmetic appertains on encrypted data without the need to decrypt is widely used to encrypt data in the cloud platform and privacy preservation for machine learning training. A recently proposed HE scheme uses nested matrices to encrypt and decrypt any field numbers of data so it can be used with machine learning and artificial intelligence applications. In this work, we propose an enhancement to the model which will increase the security and the randomness of the model and make it more reliable to be used with machine learning models. We also present a comparative analysis showing the performance of the originally suggested scheme and the improved scheme proposed in this work.

Keywords: Cryptography · Key management · Data encryption · Computations on matrices · Cryptanalysis and other attacks · Distributed systems security · Software security engineering · Privacy protections · Usability in security and privacy · Document searching · Reference works · Quantum computing

1 Introduction

Until around 2009 the idea of accessing encrypted data was just an idea. Accessing encrypted data was a demand to shield data being stored and accessed on clouds, in order to guarantee privacy as well as security of data. In 2009 [10] proposed a homomorphic encryption (HE) scheme over integers with a simple technique, which was tempting for researchers due to simplicity and promising features of security. A lattice-based implementation for the scheme by [11] was a golden ticket for the scheme to spread and get even more attention. Since it produced a homomorphic post-quantum encryption scheme. Moreover, In artificial intelligence (AI) and machine learning(ML) applications, data secrecy and privacy are major concerns standing against feeding the proper dataset to these models. Since the quantity and quality of these datasets are limited due to these concerns [1, 8, 19].

J. Abawajy et al. (Eds.): ICICCT 2023, CCIS 1841, pp. 169–181, 2023.
https://doi.org/10.1007/978-3-031-43838-7_13

Many popular encryption schemes based on the homomorphic property such as RAS and Paillier [6] were created in recent decades. Nevertheless, many researchers have used the HE property to generate or enhance new HE algorithms, such as the [13] model. The [13] is HE used to encrypt the real numbers used in machine learning applications. The encryption scheme depends on the nonabelian rings matrix and conjugacy search problem, where nested matrices with random numbers of the plaintext are used for encryption and decryption. However, the lower triangle of the matrix contains zeros indicating that randomness is not fully utilized in the lower triangle.

Therefore, in this paper, the proposed HE scheme is an enhanced algorithm of the [13] where the lower triangle of the nested matrix is utilized to increase randomness by replacing the internal zero matrices of the lower triangle of the nested matrix with message-coding. The added message-coding matrix is generated randomly from three real numbers where their summation equals the message and utilized in a nested matrix rather than zeros to increase randomness. Furthermore, the decryption algorithm uses two internal matrices located in the upper and lower triangles rather than only the upper triangle to retrieve the message from the ciphertext. Consequently, the security of the proposed HE algorithms increased without increasing the computational complexity.

This paper is organized as follows. Section 2 presents a literature review. Section 3 presents the previous and proposed HE models. Section 4 illustrated the proposed HE models. Section 5 presents the experiments. Section 6 discusses the results. Finally, Sect. 7 concludes the paper.

2 Related Work

In this section, we present a background for the methods which are proposed in the literature to enhance the privacy and secrecy of ML and AI models. Federate learning and HE are the most famous methods used for this purpose.

At the time federated learning opens the door to building ML models with secured and protected data against privacy and security intrusions. These models rely on a single central server which is prone to different forms of attacks starting from failures to backdoor attacks, and model corruption and ending with adversarial attacks and external attacks [14]. The complexity which comes with federated learning models' implementation is also considered a concern [5].

Other methods were proposed to protect the data used with ML and AI models such as differential privacy methods [20]. Multi-Party computation methods [7] and functional encryption [18] are also methods used to allow data to be used in ML and AI applications and preserve its security and privacy at the same time.

The work in [12] showed a piece of practical evidence that homomorphically encrypted data can be used efficiently with AI models and bring results close to the results produced when non-encrypted data was used with these models. In [2, 3] the authors proposed pruning techniques that allow for speeding up training ML and AI models with encrypted data. While in [13] federated learning and HE were combined to create a machine learning framework of multi-party privacy preservation strategy.

In [4] friendly CNN with HE was presented, and the work focused on creating image classification CNN which can be used for encrypted images. While the work in [13] looked outside the box and took HE into a more comfortable area for ML and AI by using nested matrices to hide the data used with these models. Meanwhile, the scheme requires less computational complexity than the other presented HE methods in the literature.

At the time HE usage in ML and AI applications suffer large computational complexity, noise accumulation, and applicability with small integer values level [17]. The wheel keeps spinning forward in the hope of creating an efficient usable HE scheme that can be used in artificial applications. We consider this work a step forward to create such a scheme, we wish to improve the scheme presented in [13] by improving its randomness and security as will be discussed soon in the next section.

3 Methodology

The proposed HE model enhances the previous HE models [13] by increasing randomness and avoiding using a sparse matrix. Consequently, the proposed HE model increases security, meanwhile, preserves computational complexity. Similar to the previous HE model, the proposed HE models have the ability to be used for multi-dimensional nested matrices. However, the proposed HE model is partially homomorphic. In this section, the previous HE model is illustrated briefly for 2×2 nested matrix in Sect. 3.1, and the proposed HE model in Sect. 3.2.

3.1 Previous HE Model

A 2×2 non-abelian matrix Ω that consists of any real numbers R is used to generate a key K, presented as 2×2 symmetric invertible matrix, and the inverse matrix of the key K^{-1} for both encryption and decryption. The encryption algorithm encrypts the message $MSG \in R$ by generating two random numbers whose summation is equal to MSG, as illustrated in Eq. (1). Then these two numbers are used to generate a symmetric invertible matrix $M1$ called message-coding as illustrated in Eq. (2). As presented in Eq. (3), the $M1$ is inserted into a nested matrix to generate ciphertext C.

$$MSG = m1 + m2 \tag{1}$$

$$M1 = \begin{pmatrix} m1 & m2 \\ m2 & m1 \end{pmatrix} \tag{2}$$

$$C = ENC(M1, K) = K \cdot \begin{pmatrix} M1 & r_1 \\ 0 & r_2 \end{pmatrix} \cdot K^{-1} \tag{3}$$

where $m1$ and $m2$ are random numbers in R, while $r1$, and $r2$ are selected randomly from Ω. Since the encryption algorithm uses a sparse matrix on the lower triangle of the nested matrix, the decryption algorithm utilizes only the upper triangle to retrieve the *MSG* from ciphertext as illustrated in Eqs. (4), (5) and (6).

$$U = DEC(C, K) = K^{-1} \cdot C \cdot K \tag{4}$$

$$\overline{U} = U_{11} \tag{5}$$

$$MSG = (\overline{U}_1)_{11} + (\overline{U}_1)_{12} \tag{6}$$

where U is the decrypted nested matrix, $\overline{U}$ is the internal matrix on the upper triangle of U. Both $(\overline{U}_1)_{11}$ and $(\overline{U}_1)_{12}$ indicates to scalar numbers in the first row of $\overline{U}$.

3.2 Proposed HE Model

The proposed model aims to increase randomness in the ciphertext of the previous HE model by replacing zeros in the lower triangle with random numbers. Furthermore, the proposed HE model enhances the decryption algorithm by utilizing both the upper and lower triangles of ciphertext to retrieve *MSG* rather than only using the upper triangle. Therefore, three random numbers are generated randomly where their summation equals *MSG* as illustrated in Eq. (7). Then, two messages-coding $M1$ and $M2$ are generated to encode *MSG*, where $M2$ is used to replace zeros in the lower triangle of ciphertexts in the previous HE model. The $M1$ and $M2$ are illustrated mathematically in Eqs. (8) and (9) respectively.

$$MSG = m1 + m2 + m3 \tag{7}$$

$$M1 = \begin{pmatrix} m1 & m2 \\ m2 & m1 \end{pmatrix} \tag{8}$$

$$M2 = \begin{pmatrix} m2 & m3 \\ m3 & m2 \end{pmatrix} \tag{9}$$

where $m1$, $m2$, and $m3$ are random numbers in R. After that, $M1, M2, K$, and K^{-1} are used to generate ciphertext C presented as 2×2 nested matrix. Equation (10) illustrates how to generate C mathematically.

$$C = ENC(M1, M2, K) = K \cdot \begin{pmatrix} M1 & r_1 \\ M2 & r_2 \end{pmatrix} \cdot K^{-1} \tag{10}$$

Equations (11), (12), and (13) demonstrate the proposed decryption algorithm that utilizes two internal matrices of decrypted nested ciphertext U to retrieve MSG. These two internal matrices are located in the upper and lower triangles of U respectively.

$$U = DEC(C, K) = K^{-1} \cdot C \cdot K \tag{11}$$

$$\overline{U} = \begin{pmatrix} U_{11} \\ U_{21} \end{pmatrix} \tag{12}$$

$$MSG = (\overline{U}_1)_{11} + (\overline{U}_1)_{12} + (\overline{U}_2)_{22} \tag{13}$$

where U is nested matrix. U_{11} and U_{21} are the internal matrices in the upper and lower triangles of U receptively. The $(\overline{U}_1)_{11}$ and $(\overline{U}_1)_{12}$ are two scalars in the first row of the U_{11}, while the $(\overline{U}_2)_{22}$ is the second scalar in the first row of the U_{21}.

4   Homomorphic Property Proofs

This section is organized as follows. The homomorphic property for 2×2 nested matrix is illustrated in Sect. 4.1. The partially homomorphic property for the proposed encrypted HE model is demonstrated in Sect. 4.2.

4.1   Homomorphic Property for 2×2 Nested Matrix

Equations (14), (15), (16), and (17) are illustrated the homomorphic property for 2×2 nested matrix, where the $C \circ \overline{C}$ indicates either additive or multiplication operations between two nested matrices C and $\overline{C}$. Since the encryption model is based on conjugate transformation by using two messages-coding in the lower and upper triangles, the decryption of the message is presented in Eq. (17). Therefore, both the addition and multiplication homomorphisms are proved using Eq. (24) in the following Sections.

$$C_o = C \circ \overline{C} = ENC(M1, M2, K) \circ ENC(\overline{M1}, \overline{M2}, K) \tag{14}$$

$$C_o = K \cdot \begin{pmatrix} M1\ r_1 \\ M2\ r_2 \end{pmatrix} \cdot K^{-1} \circ K \cdot \begin{pmatrix} \overline{M1}\ \overline{r_1} \\ \overline{M2}\ \overline{r_2} \end{pmatrix} \cdot K^{-1} \tag{15}$$

$$C_o = K \cdot \begin{pmatrix} M1 \circ \overline{M1}\ r_1 \circ \overline{r_1} \\ M2 \circ \overline{M2}\ r_2 \circ \overline{r_2} \end{pmatrix} \cdot K^{-1} \tag{16}$$

Hence,

$$DCE(C_o) = DEC(C \circ \overline{C}) = \begin{pmatrix} M1 \circ \overline{M1} \\ M2 \circ \overline{M2} \end{pmatrix} \tag{17}$$

where, $\overline{M1}$, and $\overline{M2}$ are messages-coding for another message $\overline{MSG}$. Also, $\overline{r1}$ and $\overline{r2}$ are selected randomly from Ω.

4.2 Encryption Algorithm with Partially Homomorphic Property

The HE model achieves the additional homomorphic property. However, it has not achieved the multiplication homomorphic property as presented below.

Addition Homomorphic Property: The $M\,1$ and $M\,2$ in Eqs. (8) and (9) respectively, are compensating in Eq. (17) to prove the additive homomorphic property as follows.

$$DEC(C_\oplus) = DEC\left(C + \overline{C}\right) = \left(\begin{array}{c}\left(\begin{array}{cc} m1 & m2 \\ m2 & m1 \end{array}\right) + \left(\begin{array}{cc} \overline{m1} & \overline{m2} \\ \overline{m2} & \overline{m1} \end{array}\right) \\ \left(\begin{array}{cc} m2 & m3 \\ m3 & m2 \end{array}\right) + \left(\begin{array}{cc} \overline{m2} & \overline{m3} \\ \overline{m3} & \overline{m2} \end{array}\right)\end{array}\right) \tag{18}$$

Thus,

$$DEC(C_\oplus) = \left(\begin{array}{c}\left(\begin{array}{cc} m1 + \overline{m1} & m2 + \overline{m2} \\ m2 + \overline{m2} & m1 + \overline{m1} \end{array}\right) \\ \left(\begin{array}{cc} m2 + \overline{m2} & m3 + \overline{m3} \\ m3 + \overline{m3} & m2 + \overline{m2} \end{array}\right)\end{array}\right) \tag{19}$$

Therefore,

$$DEC(C_\oplus) = \left(DEC(C_\oplus)_1\right)_{11} + \left(DEC(C_\oplus)_1\right)_{12} + \left(DEC(C_\oplus)_2\right)_{12}$$
$$= m1 + \overline{m1} + m2 + \overline{m2} + m3 + \overline{m3}$$
$$= (m1 + m2 + m3) + \left(\overline{m1} + \overline{m2} + \overline{m3}\right)$$

Since $MSG = m1 + m2 + m3$, and $\overline{MSG} = \overline{m1} + \overline{m2} + \overline{m3}$.
Hence, $DEC(C_\oplus) = MSG + \overline{MSG}$

Multiplication Homomorphic Property: Using the second message-coding $M\,2$ presented in Eq. (9), the proposed model does not achieve the fully homomorphic property because the multiplication operation is not applicable, as proved below, where the $M\,1$ and $M\,2$, as presented in Eqs. (8) and (9) respectively, are compensating in Eq. (17).

$$DEC(C_\otimes) = DEC\left(C \otimes \overline{C}\right) = \left(\begin{array}{c}\left(\begin{array}{cc} m1 & m2 \\ m2 & m1 \end{array}\right) \cdot \left(\begin{array}{cc} \overline{m1} & \overline{m2} \\ \overline{m2} & \overline{m1} \end{array}\right) \\ \left(\begin{array}{cc} m2 & m3 \\ m3 & m2 \end{array}\right) \cdot \left(\begin{array}{cc} \overline{m2} & \overline{m3} \\ \overline{m3} & \overline{m2} \end{array}\right)\end{array}\right) \tag{20}$$

Thus,

$$DEC(C_\otimes) = \left(\begin{pmatrix} m1 \cdot \overline{m1} + m2 \cdot \overline{m2} \ m1 \cdot \overline{m2} + m2 \cdot \overline{m1} \\ m2 \cdot \overline{m1} + m1 \cdot \overline{m2} \ m2 \cdot \overline{m2} + m1 \cdot \overline{m1} \end{pmatrix} \atop \begin{pmatrix} m2 \cdot \overline{m2} + m3 \cdot \overline{m3} \ m2 \cdot \overline{m3} + m3 \cdot \overline{m2} \\ m3 \cdot \overline{m2} + m2 \cdot \overline{m3} \ m3 \cdot \overline{m3} + m2 \cdot \overline{m2} \end{pmatrix} \right) \tag{21}$$

Therefore,

$$DEC(C_\oplus) = \left(DEC(C_\oplus)_1\right)_{11} + \left(DEC(C_\oplus)_1\right)_{12} + \left(DEC(C_\oplus)_2\right)_{12}$$
$$= m1 \cdot \overline{m1} + m2 \cdot \overline{m2} + m1 \cdot \overline{m2} + m2 \cdot \overline{m1} + m2 \cdot \overline{m3} + m3 \cdot \overline{m2}$$

Since $MSG = m1 + m2 + m3$, and $\overline{MSG} = \overline{m1} + \overline{m2} + \overline{m3}$.
Hence, $DEC(C_\otimes) \neq MSG \cdot \overline{MSG}$.

5 Experimental Results

5.1 Evaluation Metrics

Since each scalar in plaintext is presented as a nested matrix in ciphertext, the determinant value of each nested matrix is used to compute the scalar value of the ciphertext. Then, both ciphertext and plaintext are converted to a sequence of binary numbers consisting of zeros and ones. Three randomness test methods are used to evaluate the randomness of the proposed 2×2 nested matrices HE model [15]: Monobit frequency, constant run, and Poker tests.

The Monobit frequency is the most popular test that detects whether the zeros and ones have a uniform distribution over the entire sequence [15]. The constant run test examines the velocity of oscillation between zeros and ones in sequence by computing the count of zeros and ones runs over a total count of runs [9, 15]. The Poker test evaluates the independence of chunks consist k bits by computing the frequency of chunks having the same values of bits in the entire sequence [15]. Furthermore, the chi-square distribution is used to examine whether the tests are passed or not [16]. In the proposed experiments, the acceptable ratio of error is 0.005 with degree freedoms 1, 4, and 7 for Monobit frequency, constant run, and poker tests, respectively. Once the value of each test is less than 0.01, the test fails due to the sequence is not random.

5.2 Results

Table 1 presents the results of the Monobit frequency, run, and poker tests for both the proposed HE and previous HE models [13] for various sizes of plaintext. It shows that the ciphertexts of proposed HE has more uniform distribution than the previous HE model [13]. The frequencies of the proposed HE model are more closed to zero than frequencies in previous HE for different sizes of inputs. Although binary sequences of both models tend to contain ones more than zeros, as demonstrated in Fig. 1, the proposed model has more occurrence of ones which in turn reduces using a nested sparse matrix in ciphertext compared to the previous HE model.

Table 1. Presents the results of the Monobit frequency, run, poker tests for both the proposed HE and previous HE models [13] for various sizes of plaintext.

Size of input	Monobit frequency		Run test		Poker	
	Proposed	Previous	Proposed	Previous	Proposed	Previous
10	1.44 ∓ 1.60	1.63 ∓ 1.11	2.30 ∓ 0.85	2.43 ∓ 0.89	4.63 ∓ 1.00	4.65 ∓ 1.85
20	1.07 ∓ 1.03	1.45 ∓ 1.65	2.97 ∓ 0.81	2.74 ∓ 0.81	4.48 ∓ 1.50	5.07 ∓ 1.26
30	1.11 ∓ 0.89	1.75 ∓ 1.81	2.56 ∓ 0.61	2.81 ∓ 0.52	4.26 ∓ 1.61	4.58 ∓ 1.80
40	2.68 ∓ 2.20	2.05 ∓ 1.62	2.96 ∓ 0.70	2.22 ∓ 1.04	4.45 ∓ 1.61	5.42 ∓ 1.63
50	2.92 ∓ 2.38	3.10 ∓ 2.04	3.11 ∓ 0.62	2.83 ∓ 0.79	5.44 ∓ 0.90	5.85 ∓ 0.73
60	3.38 ∓ 2.09	2.19 ∓ 1.41	2.93 ∓ 0.68	2.85 ∓ 0.91	5.22 ∓ 1.14	5.14 ∓ 1.59
70	4.17 ∓ 1.88	1.46 ∓ 1.60	2.83 ∓ 0.82	2.89 ∓ 0.68	4.30 ∓ 1.65	4.69 ∓ 1.47
80	3.27 ∓ 2.44	3.48 ∓ 2.51	3.13 ∓ 0.64	3.09 ∓ 0.93	4.82 ∓ 1.18	5.57 ∓ 0.80
90	4.61 ∓ 2.15	2.84 ∓ 1.69	3.32 ∓ 0.62	3.29 ∓ 0.67	4.57 ∓ 1.58	4.70 ∓ 1.10
100	3.25 ∓ 2.15	3.30 ∓ 1.83	2.96 ∓ 0.77	2.92 ∓ 0.52	5.10 ∓ 1.50	5.36 ∓ 1.09

Furthermore, Table 1 illustrates that the velocity of oscillation in the proposed HE is higher than the previous HE model [13] for different sizes of inputs. Therefore, the proposed HE model is more random. Although, Table 1 presents that the proposed HE has less independence compared to the previous HE model [13], the proposed HE model has less number of test failures for the Monobit frequency, constant run, and poker tests as illustrated in Fig. 2, 3, and 4 respectively. In Fig. 2 we present the monobit frequency test results, this test measures if the cardinality of 0s and 1s produced by the generator are approximately the same; this is the case for a truly random sequence. The proposed encryption model proposed in this work shows better results which means that the randomness of the encryption process is enhanced, especially for large input sizes (Figs. 5 and 6).

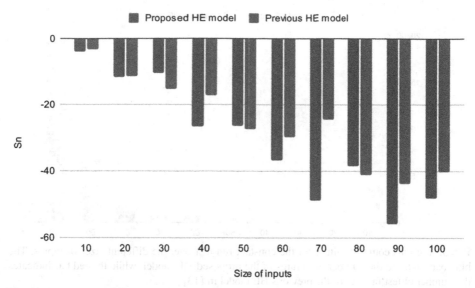

Fig. 1. Presents the differences between the count of zeros and ones in a sequence of ciphertext Sn for both proposed HE and previous HE [13]. It demonstrates that the occurrence of one is higher than the occurrence of zero.

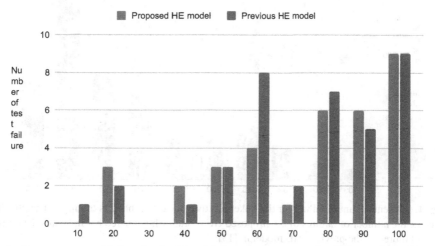

Fig. 2. Presents count of failure for the Monobit frequency test over the different sizes of inputs. The blue bars indicate the number of failures of the proposed HE model, while the red bar indicates the number of test failures of the previous HE model in [13].

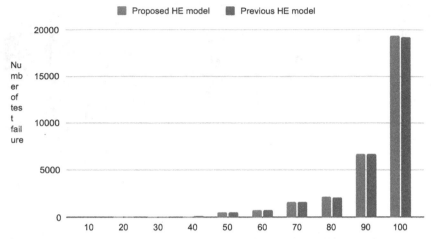

Fig. 3. Presents count of failure for the constant run test over the different sizes of inputs. The blue bars indicate the number of failures of the proposed HE model, while the red bar indicates the number of test failures of the previous HE model in [13].

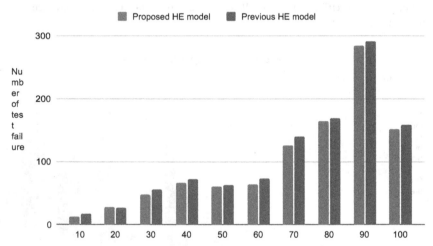

Fig. 4. Presents count of failure for the Poker test over the different sizes of inputs. The blue bars indicate the number of failures of the proposed HE model, while the red bar indicates the number of test failures of the previous HE model in [13].

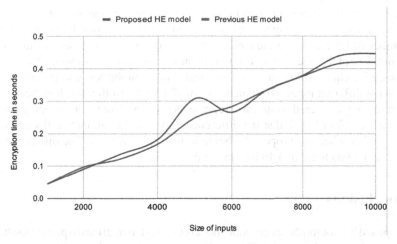

Fig. 5. Shows the computational time of the encryption algorithm for both the proposed PHE model and previous HE models.

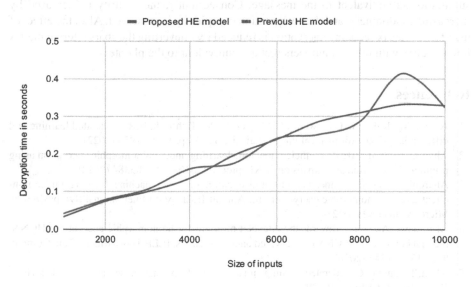

Fig. 6. Shows the computational time of the encryption algorithm for both the proposed PHE model and previous HE models

6 Discussion

The homomorphic encryption scheme presented in this work aims to increase the randomness of the encryption scheme presented in [13]. The randomness increase comes through modifying Eqs. (1), (2) and (3) to (7), (8) and (9). Increasing the randomness should increase the security of the encryption method. We compare the randomness of the encryption scheme presented in [13] to the encryption scheme presented in this work. The randomness of cipher texts is usually measured by tests that focus on comparing the number of 0 s and 1 s in the cipher; i.e. monobit test. The closer the number of 0 to the number of 1 s the better. The randomness can also be measured by the length of consecutive chains of 0 s or 1 s i.e. run test. The monobit test and the run test were measured at different input sizes ranging from 10 to 100. Both tests show better results with smaller input sizes, the input size represents the length of the array being ciphered. We can notice from Fig. 1 that the difference between the number of 0 s and ones is always smaller when the proposed scheme in this work is used. The difference grows from around 2 bits to around 40 bits unlike the original.

7 Conclusions

The proposed homomorphic encryption models use nested matrices to protect the training of the machine learning model. The randomness is increased in the lower triangle of the nested matrix for the encryption algorithm, thus the decryption algorithm utilized both upper and lower triangles of the nested matrix to retrieve the number that their summation is equivalent to the message. Consequently, the security is increased by increasing randomness, and the computation complexity is preserved. Also, the effect of using zero-matrix reducing encryption is reduced by converting the spars internal matrix into a matrix with random numbers that are equivalent to the plaintext.

References

1. Adnan, M., Kalra, S., Cresswell, J.C., Taylor, G.W., Tizhoosh, H.R.: Federated learning and differential privacy for medical image analysis. Sci. Rep. **12**(1), 1953 (2022)
2. Aharoni, E., et al.: HE-PEx: efficient machine learning under homomorphic encryption using pruning, permutation and expansion. arXiv preprint arXiv:2207.03384 (2022)
3. Aharoni, E., et al.: Prune, permute and expand: efficient machine learning under non-client-aided homomorphic encryption. In: Annual IEEE/ACM International Symposium on Microarchitecture (2022)
4. Al Badawi, A., et al.: Towards the AlexNet moment for homomorphic encryption: HCNN, the first homomorphic CNN on encrypted data with GPUs. IEEE Trans. Emerg. Top. Comput. **9**(3), 1330–1343 (2020)
5. Alam, T., Gupta, R.: Federated learning and its role in the privacy preservation of IoT devices. Future Internet **14**(9), 246 (2022)
6. Alaya, B., Laouamer, L., Msilini, N.: Homomorphic encryption systems statement: trends and challenges. Comput. Sci. Rev. **36**, 100235 (2020)
7. Berry, C., Komninos, N.: Efficient optimisation framework for convolutional neural networks with secure multiparty computation. Comput. Secur. **117**, 102679 (2022)
8. Dong, T., Zhao, B., Lyu, L.: Privacy for free: how does dataset condensation help privacy? In: International Conference on Machine Learning, pp. 5378–5396. PMLR (2022)
9. Dragomir, I.R., et al.: Statistical assessment of binary sequences generated by cryptographic algorithms. Dezbateri Soc. Econ. **5**(2), 23–31 (2016)
10. Gentry, C.: Fully homomorphic encryption using ideal lattices. In: Proceedings of the Forty-First Annual ACM Symposium on Theory of Computing, pp. 169–178 (2009)

11. Gentry, C., Halevi, S.: Implementing gentry's fully-homomorphic encryption scheme. In: Paterson, K.G. (eds.) EUROCRYPT 2011. LNCS, vol. 6632, pp. 129–148. Springer, Heidelberg (2011). https://doi.org/10.1007/978-3-642-20465-4_9
12. Lee, J.W., et al.: Privacy-preserving machine learning with fully homomorphic encryption for deep neural network. IEEE Access **10**, 30039–30054 (2022)
13. Li, J., Kuang, X., Lin, S., Ma, X., Tang, Y.: Privacy preservation for machine learning training and classification based on homomorphic encryption schemes. Inf. Sci. **526**, 166–179 (2020)
14. Lian, Z., Su, C.: Decentralized federated learning for Internet of Things anomaly detection. In: Proceedings of the 2022 ACM on Asia Conference on Computer and Communications Security, pp. 1249–1251 (2022)
15. Mengdi, Z., Xiaojuan, Z., Yayun, Z., Siwei, M.: Overview of randomness test on cryptographic algorithms. J. Phys. Conf. Ser. **1861**, 012009 (2021)
16. Nam, J.W., Kim, J., Hong, J.P.: Stochastic cell-and bit-discard technique to improve randomness of a TRNG. Electronics **11**(11), 1735 (2022)
17. Popescu, A.B., et al.: Privacy preserving classification of EEG data using machine learning and homomorphic encryption. Appl. Sci. **11**(16), 7360 (2021)
18. Shahzad, K., Zia, T., Qazi, E.u.H.: A review of functional encryption in IoT applications. Sensors **22**(19), 7567 (2022)
19. White, T., Blok, E., Calhoun, V.D.: Data sharing and privacy issues in neuroimaging research: opportunities, obstacles, challenges, and monsters under the bed. Hum. Brain Mapp. **43**(1), 278–291 (2022)
20. Zhao, Y., Chen, J.: A survey on differential privacy for unstructured data content. ACM Comput. Surv. (CSUR) **54**(10s), 1–28 (2022)

A Deep Learning Model for Heterogeneous Dataset Analysis - Application to Winter Wheat Crop Yield Prediction

Yogesh Bansal[1]([✉]) [iD], David Lillis[1] [iD], and M.-Tahar Kechadi[1,2] [iD]

[1] School of Computer Science, University College Dublin, Dublin, Ireland
yogesh.bansal@ucdconnect.ie, {david.lillis,tahar.kechadi}@ucd.ie
[2] Insight Centre for Data Analytics, Dublin, Ireland

Abstract. Western countries rely heavily on wheat, and yield prediction is crucial. Time-series deep learning models, such as Long Short Term Memory (LSTM), have already been explored and applied to yield prediction. Existing literature reports that they perform better than traditional Machine Learning (ML) models. However, the existing LSTM cannot handle heterogeneous datasets (a combination of data that varies and remains static with time). In this paper, we propose an efficient deep learning model that can deal with heterogeneous datasets. We developed the system architecture and applied it to the real-world dataset in the digital agriculture area. We showed that it outperformed the existing ML models.

Keywords: Deep-Learning Model · Digital Agriculture · Heterogeneous Time-series Dataset · Machine Learning models · Winter Wheat Crop Yield Prediction

1 Introduction

Crop yield prediction is a nonlinear process and usually involves analysing several features coming from multiple heterogeneous datasets. Time series deep learning models like LSTM [18, 19] are proven to be good models in case of time series data however, when dealing with mixed type datasets like soil and weather, integrating them into an LSTM model can be challenging. Also, heterogeneous datasets may contain different types of data which would require different pre-processing methods to be effectively used in a LSTM model. It is designed specifically to work well only for time series datasets, however there is a need to build a system architecture into it to incorporate the ability of handling heterogeneous datasets more effectively.

In a LSTM model, various optimisation techniques such as gradient descent, momentum, and Adam are most commonly used in neural networks to minimise the error. Gradient descent [5, 18, 19, 25] is an optimisation technique which measures the change in weights and finds the parameter values to minimise a cost function. Another approach momentum [5] incorporates the information about gradient's previous direction and accumulates the gradient values over time to accelerate optimisation by introducing a

J. Abawajy et al. (Eds.): ICICCT 2023, CCIS 1841, pp. 182–194, 2023.
https://doi.org/10.1007/978-3-031-43838-7_14

new variable momentum. The Adam approach [5, 11, 21, 23, 25] combines two parameters: momentum and adaptive learning rate. It adjusts the learning rate dynamically based on the loss function gradient.

While using time series deep learning models, hyper-parameters play a major role and should be tuned for optimal performance of the model. The hyperparameters used in this study are epochs, learning rate, the number of hidden units. Selecting appropriate hyper-parameters' values is not straightforward. It is usually based on trial and error. The LSTM optimal hyper-parameter values depend on the specific task, dataset, complexity of the model and the amount of training data available. In this paper, we develop a system architecture to modify the existing LSTM model to be able to handle both time-series and non times-series data effectively.

The paper is structured as follows: Literature is presented in Sect. 2, followed by data description in Sect. 3. Pre-processing information is provided in Sect. 4, followed by "experiment setup" and "results" in Sect. 6 and 7, respectively. Section 8 summarizes the study.

2 Literature

Data-driven agricultural crop yield utilizing several types of datasets, including soil and weather has been addressed by many researchers, mainly from the agriculture point of view. Therefore, we focus on the studies related to crop yield predictions using ML including deep learning approaches [3, 17, 22]. Hybrid models which are a combination of ML models have been proven to perform better than individual ML models for crop yield predictions [21, 23]. The most commonly used deep learning models are the convolutional neural networks (CNN) and LSTM. Mathieu et al. [12] assessed agro-climatic indices over medium and low production areas. They concluded that agro-climatic indices can significantly improve crop yield modelling in comparison with direct weather variables and highlighted temperature and precipitation as the most crucial weather factors affecting crop yield. Agro-climatic indices usually provide guidance on the types of crops that are best suited for a particular region, as well as the optimal times for planting and harvesting those crops. This is very important in agriculture because by understanding the agro-climatic conditions of a particular region, farmers can better manage their crops and improve their yields while minimising the risks associated with climate variability.

A full review of the use of machine learning models to crop management can be found in [6]. In [10], a random forest (RF) model was used to predict wheat production, with multiple linear regression (MLR) serving as the standard of comparison. All performance data indicate that RF outperforms MLR. A similar study conducted by [16] compared RF, XGBoost, and KNN for crop yield prediction on rainfall and temperature data and found that RF performed the best. Another study [23] use ML including neural models to estimate winter wheat production on soil and weather data and found that neural models outperformed other models. [7] conducted more relevant research, indicating that RF is a superior ML predictor than other ML models.

Other studies [1, 8] utilised ensemble ML models and compared them to a single ML model and noted that ensemble ML models align more towards making predictions close to actual yield values. [18] introduces a model for performing in-season soybean yield predictions utilizing Long-Short Term Memory (LSTM) and traditional ML models on weather and satellite data.

Sun et al. [21] proposed a deep CNN-LSTM model for both end-of-season and in-season soybean yield prediction based on county-level weather data. The proposed CNN-LSTM model outperformed either the CNN or LSTM model in terms of prediction performance. They assert that this form of model architecture may significantly enhance yield prediction for more crops, including corn, wheat, and potatoes. In a similar study representing hybrid models, [23] developed a LSTM-CNN to estimate winter wheat yield at the county level in Chin using weather and remote sensing data. Results demonstrated that LSTM-CNN enhanced the model's yield prediction ability. In another study [11], an LSTM model is built that combines crop phenology, weather, and remote sensing data to predict county-level corn yields. The results showed that LSTM model outperformed other ML methods for estimating end-of-season yield.

In one of the studies conducted in India [19], an LSTM model was proposed to estimate agricultural yields using satellite data at the block level across many states. The proposed strategy surpassed classical ML approaches by more than 50%. They also demonstrate that the incorporation of contextual information, such as the location of farms, water sources, and metropolitan areas, improves yield estimations.

[5] presented a more accurate optimiser function (IOF) and used it along with LSTM model. The proposed model is compared to NN, RNN, and LSTM, and the results indicate that the proposed IOF minimises training error by addressing under-fitting and overfitting. The findings indicate that the recommended IOFLSTM has the benefit of accurate crop yield prediction. The decrease in RMSE for the proposed model implies that the proposed IOFLSTM can beat the CNN, RNN, and LSTM when predicting crop yield.

Several strategies for estimating crop yields using soil and environmental characteristics was explored in [9]. It compared LSTM with other ML models and concluded that LSTM is effective in comparison to the others for all the evaluation metrics used in the study. [2] presented the RNN-LSTM model to estimate the wheat crop yield of India's northern area using a 43-year benchmark dataset and compared it with artificial neural networks, RF, and multivariate linear regression. The performance of RNN-LSTM model came out to be much better than the other models utilised.

Another combination of LSTM-RF framework was introduced in [20] for forecasting wheat yield using vegetation indicators and canopy water stress indices at several growth stages. In comparison to LSTM, the LSTM-RF model produced more accurate predictions. The findings demonstrated that LSTM-RF evaluated both the time-series features of winter wheat development and the nonlinear characteristics between remote sensing data and crop yield data, therefore offering an alternate method for yield prediction in contemporary agricultural production. Regarding data pre-processing, Ngo et al. [14] utilised neighbouring fields to fill in missing values for soil properties such as pH. Bansal et al. [4] indicated the importance of weather data in crop yield predictions. They demonstrated statistically that the addition of weather data to soil data improves yield prediction. This study is considered as the benchmark study because of the same dataset and same train/test splits for training and testing. The experimental results obtained from the proposed models are compared with the best-performing ML model in this study [4].

The trend from the literature showed that for mixed-type datasets that have geographical and temporal dimensions, researchers rely on the hybrid models i.e., combining time-series model with either traditional ML or neural models but there does not exist any model in the literature which could effectively handle mixed-type datasets with geographical and temporal dimensions.

3 Data Description

The datasets utilised include soil and weather information for numerous farms across several years. Multiple fields make up a farm, and each field is further subdivided into zones. In the context of crop management, a "zone" refers to a sub-region within a field [22]. Let Z be a set of zones: $Z = \{z_1, z_2, ..., z_n\}$. For each zone z_i by year, the following features are grouped into two categories: soil data and weather data. The soil data has been collected from various farms along with the weather data for those farms for 6 years i.e., 2013–2018. Moreover, we have done significant work in organising and integrating the datasets we have collected so far, and these were reported in [13, 15].

The data on the soil comprises information regarding the results of soil testing carried out in the agricultural zones. Because of their high cost and the fact that the values do not change much over relatively short spans of time, these soil tests are only performed on a very rare basis (usually every three to four years). As we are just examining one year's worth of data for each instance, we will assume that the soil's attributes will not vary dramatically over the course of this time period and hence treat them as constants. In the event that a zone does not have a soil test in a particular year, the soil is mapped based on the results of the most recent test that was performed in the same zone the year before. Data about the weather, on the other hand, is a collection of different points in time. An instance I in the dataset can be represented as follows:

$$I = [Z, S, W, T, Y] \tag{1}$$

where $S = \{s_1, s_2, \cdots, s_x\}$ is the set of x soil variables. $W = \{w_1, w_2, \cdots, w_y\}$ is the set of y weather variables, T is time in weeks, and Y is yield.

In this study soil and weather data characteristics cover the period from 2013 to 2018. It includes soil nutrients (P, K, Mg), physical features (soil type, stone content), chemical properties (organic matter, $CaCO_3$, pH), yield for a zone, and sowing and harvesting dates. The multiple soil type classifications in the dataset include shallow, medium, deep clay, and deep fertile; stone content is stoneless, low, moderate, and high; organic matter is low, moderate, and extremely high; and $CaCO_3$ is slightly calcium, medium calcium, high calcium, and acidic. The weather data includes air temperature, precipitation, solar radiation, and humidity.

4 Data Pre-processing

The original soil and weather data contain discrepancies, noise, missing values, and errors. Identifying and correcting data errors, employing feature engineering to extract new features [24] from weather data, integrating mixed-type soil and weather datasets, mapping soil categorical variables to numerical variables, and filtering the integrated data from the growth period till the end of the season to make it model-ready are all components of the data preparation process. Figure 1 illustrates the pre-processing steps used for soil and weather data.

In the case of soil data, we begin with crop, farm, and zone datasets containing zone yield information. These several datasets have been pre-processed and integrated to create a master dataset on the soil. Winter wheat was taken from the original soil dataset, which included a variety of other crops. Several duplicate soil testing and yield values for the same zone and year were removed. Many soil nutrient and yield data values were identified to be erroneous or unattainable with the aid of agricultural domain experts. Those are almost certainly due to human mistakes. These were eliminated from the data.

The mapping of soil categorical variables i.e., soil type, stone content, organic matter, and $CaCO_3$ is also done to numerical variables for modelling. After pre-processing the crop, farm, and zone information, they are integrated to create a pre-processed soil dataset. The principal component analysis (PCA) is applied to the pre-processed soil dataset to reduce its dimensionality and further used as a bias by the proposed deep learning model. The motivation behind using PCA is that due to the heterogeneity and a large number of features, it extracts the features that can be used to train our proposed deep learning models in an efficient and compact manner. Moreover, the extracted features are uncorrelated which are useful in eliminating multicollinearity issues in the data, and help in improving the stability and performance of the deep learning model. A bias is a parameter that is added to the inputs and the LSTM internal state to allow the network to better capture the relationship between inputs and outputs.

The weather variables i.e., air temperature, precipitation, solar radiation, and humidity are represented into weekly based features i.e., each week of all years studied (2013–2018) according to the dynamic sowing and harvest dates of zones/year. After pre-processing of weather data and extracting Two Agroclimatic indices i.e., degree days, and effective growing days are fetched from air temperature weather variables and, used with other weather attributes i.e., average temperature, accumulated precipitation, accumulated solar radiation, and average humidity.

Degree days [12] are calculated as the maximum of 0 and the average of maximum and minimum daily temperature, summed over a week. The total number of days in a week when the average temperature is greater than 5 °C is referred to as effective growing days [12]. The average temperature is calculated by averaging the temperature (T) over a week. The accumulated precipitation is calculated by summing precipitation (P) over a week. Solar radiation is summed over a week to get accumulated solar radiation. Humidity (H) is averaged over the week to get average humidity.

After soil and weather data have been pre-processed according to weekly based representation, data integration of weather and soil begins.

Figure 2 illustrates the weekly-based representation i.e., 1st week, 2nd week, ... , t^{th} week of weather data and soil data which is used as a bias for a particular zone and year). This integrated data is further filtered from the growth period of a zone i.e., from 17th week till the harvesting period of that zone.

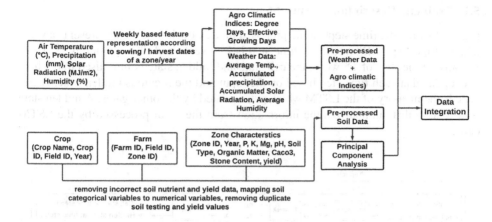

Fig. 1. Data pre-processing steps.

Zones	Year	Week	Weather (Sequence Data)			Soil (Non Sequence Data)	Yield
			Degree Days	.....	Avg Humidity		
nth zone	...	1st week	...	...	...	Principaled component used as a Bias for this zone/year	nth yield
		2nd week	...	...	...		
		....	...	...	...		
		tth week	...	...	...		

Fig. 2. Weekly-based data representation after data pre-processing.

5 Proposed Approach

This study presents a novel approach to handling both sequence and non-sequence data at the same time. We develop a deep learning model to make yield predictions on multiple zones which have different sowing and harvest dates. An epoch is complete when the proposed model in this study is trained on every zone in the dataset exactly once. The applications of classical LSTM are constrained by the requirement that it should have a sequence, i.e., time-series data. However, when we have a mixture of sequence and non-sequence data, a variation is required. Thus, we propose a variant of the classical LSTM that is capable of processing both sequence and non-sequence data. The following proposed approach entails a description of time steps, forward and backward propagation in training, and the steps in testing.

5.1 Training: Description of Time Steps

Figure 3 shows the time steps in the proposed model. Each time step represents a week i.e., 1st time step is 1st week; 2nd time step is 2nd week; and t^{th} time step is t^{th} week. Initially, at the start of 1st week, the cell state and hidden state is initialised to 0 which gets updated at each time step based on the input and the previous hidden state. A cell state is the memory of the LSTM which gets updated by the other gates. A hidden state is an output that summarizes the information from the input processed by the LSTM cells.

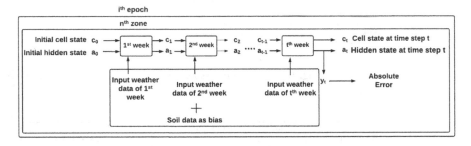

Fig. 3. Training Steps (Description of time steps)

Each zone has different weeks because of different sowing and harvesting dates which means the proposed model is trained according to the weeks present in a particular zone. In the forward time step, weather data for each respective week along with bias which is constant over the time steps for a zone is propagated from 1st week to t^{th} week of that zone. After t time steps, a yield prediction is computed which is a real number, $y \in R$. It is then compared with the actual yield to find an absolute error for that zone.

In the backward time step, gradients of weight parameters related to only sequence data i.e., weather data are generated. Here, the parameters are the weights of the network, while the gradients are the derivatives of the loss function with respect to these parameters. Since non-sequence data i.e., the soil remains constant throughout the year for a zone, it is considered as a bias and its gradients are not computed.

5.2 Training: Description of Forward/Backward Propagation

Figure 4 illustrates the forward and backward propagation steps of a proposed deep learning model. It is similar to LSTM architecture except where only gradients of sequence weather data are computed and soil Data is inputted as bias. This figure illustrates one LSTM cell, which corresponds to one zone in one week. Below are the notations used:

- f_t, u_t, and o_t represents forget, update and output gate respectively. $^R f_t$, $^R u_t$, and $^R o_t$ are the respective derivatives. A forget gate determines which information from the previous cell state should be discarded. An update gate controls how much information from the candidate state should be added to the cell state. An output gate determines which information from the current cell state should be output.

- w_{fa}, w_{ua}, and w_{oa} are the weight parameters related to the hidden state of forget, update and output gate respectively. $^{R}w_{fa}$, $^{R}w_{ua}$, $^{R}w_{oa}$ are the respective derivatives.
- w_{fx}, w_{ux}, and w_{ox} are the weight parameters related to weather data of forget, update and output gate respectively. $^{R}w_{fx}$, $^{R}w_{ux}$, $^{R}w_{ox}$ are the respective derivatives.
- a_{t-1}, a_t represents previous and current hidden state. Similarly, c_{t-1}, c_t represents previous and current cell state. $^{R}a_{t-1}$, $^{R}a_t$, $^{R}c_{t-1}$, $^{R}c_t$ are the respective derivatives.

The description of Forward/Backward Propagation is as follows:-

- In forward propagation, weight parameters of forget gate, update gate, candidate state, output gate and input comprising of weather and soil data are propagated. Here, a candidate state updates the cell state based on the input and the previous hidden state.
- After each time step/week/LSTM cell, cell state and hidden state are updated.
- Calculate the gradients of sequence weather data only by back propagation through time at time step t using the chain rule. No gradients of soil data are computed during back propagation.

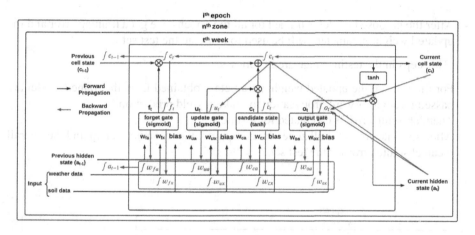

Fig. 4. Training (Description of Forward/Backward Propagation)

Training and Testing. Figure 5 shows the training and testing steps of the proposed approach.

The steps in the training stage are as follows:

- Firstly, we set the seed and randomly initialize the weight parameters. The seed is set to ensure that the generated random numbers are reproducible.
- Start the outer loop for the number of epochs the proposed model is trained.

  • For each epoch, there is an inner loop for the number of zones present in the training set. For each zone, do the following:-

- Do a forward propagation using available parameters from 1st week to t^{th} week to generate yield prediction.
- Compute the absolute error.
- Do a backward propagation to generate the gradients of weight parameters of different gates. These gradients are the derivatives of the loss function with respect to each of the parameters in the LSTM network.
- The gradients obtained from the backward propagation are based only on the sequence weather data and are used in the optimisation algorithm to adjust the model parameters during training to minimise the loss function and improve the model's ability to make accurate predictions on new data. This is where the proposed deep learning model differs from classical LSTM where there was no way to segregate the gradients of heterogeneous data.
- Absolute errors for all the zones are computed.

- After the completion of an epoch, mean absolute error is computed and the updated weight parameters are passed from the last zone of the previous epoch to the first zone of the next epoch.

– After the proposed model is trained for all the epochs, we get a training error and the updated weight parameters will be used to evaluate the test set.

The steps in the testing stage are as follows:

– For the test set, the updated weight parameters obtained from the trained model are passed through forward propagation to make a yield prediction.
– Then, absolute error is calculated for a zone.
– After absolute errors for all zones are computed, the mean is taken to find the overall mean absolute error for the test set.

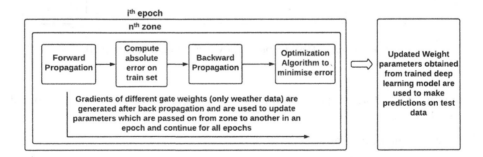

Fig. 5. Training and Testing Steps

6 Experiment Setup

The data utilised in this study covers the years 2013 to 2018. We train the proposed deep learning models from 2013 to 2017 and then test it on 2018 dataset. MAE is used as the evaluation metric.

To compare proposed deep learning models with the traditional ML models used in this study [4], we present the weather data in the same two forms: common weather attributes (i.e., temperature, humidity, precipitation, solar radiation), and agro-climatic indices (i.e., degree days, effective growing days). Also, we consider only the growth period weeks beginning from week 17 till the end of the season for modelling.

We have developed 3 deep learning models using the proposed approach which are independent of each other. The first model is using gradient descent [18] optimiser. Second model is using momentum [5] optimiser, and the third model is using Adam [21] optimiser to predict winter wheat yield.

The learning rate is a hyper-parameter that determines the step size while moving towards a minimum of a loss function during the training of a model.

During the training process, the goal is to optimize the model parameters to minimize the error between the predicted yield and the actual yield. Hidden units are the intermediate computational units within a neural network that are responsible for processing and transforming the input data into useful representations that can be used to make predictions. Hidden units are the nodes within the neural network that receive input from the previous layer and perform a computation on that input. More hidden units can enable the network to capture more complex patterns in the data, but may also increase the risk of over-fitting, while fewer hidden units may lead to underfitting and poor performance.

To find the best hyper-parameter values for our problem of yield prediction, we performed an exhaustive hyper-parameter search for each model separately using these specific values of hyper-parameters i.e., hidden units from 10, 20, 30, 40; learning rate from 0.001, 0.005, 0.01, 0.05; epochs from 10, 20, 30, 40, 50, 60, 70, 80, 90, 100; Though the optimal hyper-parameter values for an LSTM model depend on the specific task and dataset, however, these are the most commonly used by the authors in the literature who used LSTM on their specific datasets.

For each of these models separately, the proposed approach is applied to the training dataset to get a trained deep learning model and the updated parameters obtained from a trained model are used to make a prediction on the test set.

A comparison is made between the proposed deep learning models and the best-performing ML model i.e., gradient boosting in this study [4]. In addition, the proposed deep learning models are compared among themselves to determine which provides the most accurate yield prediction performance. Statistical significance tests are also done to evaluate if the differences between the MAE values of proposed deep learning models and gradient boosting are significant or not.

7 Experimental Results

This section presents the results for three proposed deep learning models. a) Proposed Deep Learning Model using Gradient Descent optimiser with best hyperparameters, b) Proposed Deep Learning Model using Momentum optimiser with best hyper-parameters c) Proposed Deep Learning Model using Adam optimiser with best hyper-parameters.

Table 1 shows the MAE of yield prediction in t/h by three proposed models developed in this study and the best performing traditional ML model i.e., Gradient Boosting from the baseline study [4]. Alongside, it also shows the statistical significance tests to measure whether the proposed models improve yield prediction over baseline, our alternative hypothesis is that the MAE of the proposed models is less than the MAE of gradient boosting in the baseline study. Here, p-values are calculated using a one-tailed paired t-test by comparing the absolute errors from a) proposed deep learning models and b) gradient boosting.

It is noted that each of the proposed deep learning models surpasses the yield prediction performance of Gradient Boosting. The best set of hyper-parameter values in the first proposed model using gradient descent optimiser is found to be $h = 20$, $lr = 40$, $e = 0.01$. It gives an MAE of 1.31 t/h whereas gradient boosting gives an MAE of 1.48 t/h.

Table 1. Winter wheat yield predictions comparison of proposed deep learning model and best ML model from baseline study [4]. Best Hyper-parameters (h = Hidden units, lr = Learning rate, e = Epochs).

Gradient Boosting Model (MAE) - Baseline [4]	Proposed Deep Learning Model (MAE)		
	Gradient Descent	Momentum	Adam
	$h = 20, lr = 0.01$	$h = 10, lr = 0.005$	$h = 10, lr = 0.005$
	$e = 40$	$e = 40$	$e = 20$
1.48	1.31	1.36	**1.22**
p value	$3.2e^{-5}$	$5.5e^{-6}$	$9e-4$

For the second proposed model using momentum optimiser, best hyperparameters are found to be $h = 10$, $lr = 40$, $e = 0.005$ and it gives an MAE of 1.36 t/h whereas gradient boosting gives an MAE of 1.48 t/h.

The third proposed model using Adam optimiser gives the least MAE of all proposed models with 1.22 t/h with its best hyper-parameters as $h = 10$, $lr = 20$, $e = 0.005$ whereas gradient boosting gives an MAE of 1.48 t/h.

It is noted from the statistical significance tests that p-values for all the proposed models are below the 5% threshold, and thus the alternative hypothesis can be accepted. Consequently, it can be said that yield prediction performances of proposed deep learning models are better than traditional ML models i.e., gradient boosting in the baseline study [4].

8 Conclusion

The purpose of the proposed deep learning models is to predict winter wheat crop yield and determine whether it performs better than the best performing ML model (i.e., Gradient Boosting in the baseline study [4]). We showed that the proposed hybrid approach is able to make better yield predictions than Gradient Boosting. Among the three proposed deep learning models, the hybrid LSTM-Adam model returned the least MAE of 1.22 t/h on the test set, while the gradient Boosting model returned an MAE of 1.48 t/h. Future work can benefit from adding more feature-engineered agro-climatic indices and robust deep learning models.

Acknowledgement. This research is funded under the SFI Strategic Partnerships Programme (16/SPP/3296) and is co-funded by Origin Enterprises Plc.

References

1. Balakrishnan, N., Muthukumarasamy, G.: Crop productionensemble machine learning model for prediction. Int. J. Comput. Sci. Softw. Eng. **5**(7), 148 (2016)
2. Bali, N., Singla, A.: Deep learning based wheat crop yield prediction model in Punjab region of North India. Appl. Artif. Intell. **35**(15), 1304–1328 (2021)
3. Bali, N., Singla, A.: Emerging trends in machine learning to predict crop yield and study its influential factors: a survey. Arch. Comput. Methods Eng. **29**(1), 95–112 (2022)
4. Bansal, Y., Lillis, D., Kechadi, T.: Winter wheat crop yield prediction on multiple heterogeneous datasets using machine learning. In: 2022 International Conference on Computational Science and Computational Intelligence (CSCI 2022), December 2022
5. Bhimavarapu, U., Battineni, G., Chintalapudi, N.: Improved optimization algorithm in LSTM to predict crop yield. Computers **12**(1), 10 (2023)
6. Chergui, N., Kechadi, M.T.: Data analytics for crop management: a big data view. J. Big Data **9**(1), 1–37 (2022)
7. Han, J., et al.: Prediction of winter wheat yield based on multi-source data and machine learning in china. Remote Sens. **12**(2), 236 (2020)
8. Huang, B.Q., Du, C.J., Zhang, Y.B., Tahar Kechadi, M.: A hybrid HMM-SVM method for online handwriting symbol recognition. In: International Conference on Intelligent Systems Design and Applications, vol. 3, pp. 887–891. IEEE Computer Society (2006)
9. Iniyan, S., Akhil Varma, V., Teja Naidu, Ch.: Crop yield prediction using machine learning techniques. Adv. Eng. Softw. **175**, 103326 (2023)
10. Jeong, J.H., et al.: Random forests for global and regional crop yield predictions. PLoS ONE **11**(6), e0156571 (2016)
11. Jiang, H., et al.: A deep learning approach to conflating heterogeneous geospatial data for corn yield estimation: a case study of the us corn belt at the county level. Glob. Change Biol. **26**(3), 1754–1766 (2020)
12. Mathieu, J.A., Aires, F.: Assessment of the agro-climatic indices to improve crop yield forecasting. Agric. For. Meteorol. **253**, 15–30 (2018)
13. Ngo, Q.H., Kechadi, T., Le-Khac, N.-A.: Knowledge representation in digital agriculture: a step towards standardised model. Comput. Electron. Agric. **199**, 107127 (2022)
14. Ngo, Q.H., Le-Khac, NA., Kechadi, T.: Predicting soil pH by using nearest fields. In: Bramer, M., Petridis, M. (eds.) SGAI 2019. LNCS, vol. 11927, pp. 480–486. Springer, Cham (2019). https://doi.org/10.1007/978-3-030-34885-4_40

15. Ngo, V.M., Kechadi, M.-T.: Electronic farming records–a framework for normalising agronomic knowledge discovery. Comput. Electron. Agric. **184**, 106074 (2021)
16. Nigam, A., Garg, S., Agrawal, A., Agrawal, P.: Crop yield prediction using machine learning algorithms. In: 2019 Fifth International Conference on Image Information Processing (ICIIP), pp. 125–130. IEEE (2019)
17. Oikonomidis, A., Catal, C., Kassahun, A.: Deep learning for crop yield prediction: a systematic literature review. N. Z. J. Crop Hortic. Sci., 1–26 (2022)
18. Schwalbert, R.A., Amado, T., Corassa, G., Pott, L.P., Vara Prasad, P.V., Ciampitti, I.A.: Satellite-based soybean yield forecast: integrating machine learning and weather data for improving crop yield prediction in southern Brazil. Agric. For. Meteorol. **284**, 107886 (2020)
19. Sharma, S., Rai, S., Krishnan, N.C.: Wheat crop yield prediction using deep LSTM model. arXiv preprint arXiv:2011.01498 (2020)
20. Shen, Y., et al.: Improving wheat yield prediction accuracy using LSTM-RF framework based on UAV thermal infrared and multispectral imagery. Agriculture **12**(6), 892 (2022)
21. Sun, J., Di, L., Sun, Z., Shen, Y., Lai, Z.: County-level soybean yield prediction using deep CNN-LSTM model. Sensors **19**(20), 4363 (2019)
22. Van Klompenburg, T., Kassahun, A., Catal, C.: Crop yield prediction using machine learning: a systematic literature review. Comput. Electron. Agric. **177**, 105709 (2020)
23. Wang, X., Huang, J., Feng, Q., Yin, D.: Winter wheat yield prediction at county level and uncertainty analysis in main wheat-producing regions of china with deep learning approaches. Remote Sens. **12**(11), 1744 (2020)
24. Wu, B., Chen, C., Kechadi, T.M., Sun, L.: A comparative evaluation of filter-based feature selection methods for hyper-spectral band selection. Int. J. Remote Sens. **34**(22), 7974–7990 (2013)
25. Zhang, Z.: Improved Adam optimizer for deep neural networks. In: 2018 IEEE/ACM 26th International Symposium on Quality of Service (IWQoS), pp. 1–2. IEEE (2018)

Correction to: Information, Communication and Computing Technology

Jemal Abawajy, João Manuel R.S. Tavares⬤, Latika Kharb⬤, Deepak Chahal⬤, and Ali Bou Nassif

Correction to:
J. Abawajy et al. (Eds.): *Information, Communication and Computing Technology*, **CCIS 1841,**
https://doi.org/10.1007/978-3-031-43838-7

The original version of this book editor's name had not been displayed correctly. This has been corrected.

The updated version of this book can be found at
https://doi.org/10.1007/978-3-031-43838-7

Author Index

J. Abawajy et al. (Eds.): ICICCT 2023, CCIS 1841, pp. 195–196, 2023.
https://doi.org/10.1007/978-3-031-43838-7

Printed in the United States
by Baker & Taylor Publisher Services